T0192829

Modelling Supply Chain Dynamics

Jose M. Framinan

Modelling Supply Chain Dynamics

Jose M. Framinan
Industrial Management School of Engineering
University of Seville
Seville, Spain

ISBN 978-3-030-79191-9 ISBN 978-3-030-79189-6 (eBook)
https://doi.org/10.1007/978-3-030-79189-6

This Springer imprint is published by the registered company Springer Nature Switzerland AG
The registered company address is: Gewerbestrasse 11, 6330 Cham, Switzerland

To my parents.

Preface

In recent years, business trends such as the globalisation process and the increase of competition across many industrial sectors have forced companies to concentrate on their core competences and to outsource those activities in which they do not excel. As a consequence, companies no longer produce and distribute their goods in isolation, but being part of a supply chain or supply network, i.e. a set of interrelated companies who ultimately deliver the goods and services to the final customer. Nevertheless, despite the prevalence of supply chains as the primary form of production and distribution, their performance can be seriously hampered by the complex dynamics resulting from the collaboration and coordination (or lack thereof) among their members. As a result, discovering, understanding and mitigating these noxious effects have been a major concern among supply chain management professionals and academics, so the topic has evolved over the years to become a key area within the field.

This book is aimed at providing the reader with the basic modelling foundations to understand and analyse the dynamic effects induced in supply chains, most notably the bullwhip effect. It assembles seminal works on supply chain models and recent developments on the topic in order to provide a unified vision of the field for researchers and practitioners who wish to grasp one of the main challenges of supply chain management.

Given the focussed nature of the book, general topics in supply chain management are kept at a minimum, with the exception perhaps of Chap. 1 where more general issues are discussed. Therefore, the reader should not expect a proper coverage of topics such as inventory management, transportation and demand forecasting, which are treated here in a very light manner. Something similar happens with the mathematics and statistics involved in the models, although in this case several appendices have been included in order to make the book (almost) self-sufficient to understand the models developed. Nevertheless, given the extension of the field—Scopus search with the keyword "bullwhip effect" shows more than 1,400 records, with almost 400 in the last 5 years—it is not possible to present a comprehensive, detailed discussion of all the topics regarding the bullwhip effect. Instead, the main results are presented, and at the end of each chapter, it is included a section with further readings with many references.

Organisation of the Book

The book does not necessarily have to be read in a sequential order, although this would be perhaps the recommended course for those approaching the topic for the first time. It is structured into nine chapters and three appendices. In the first chapter, a general introduction to the supply chains and to the bullwhip effect is given. The activities in supply chain management that most influence the bullwhip effect are presented in Chap. 2. In Chap. 3, we discuss in-depth the bullwhip effect, recollecting his 'history' and pool of causes. Different indicators to measure the dynamics of the supply chain, both at the company level and a supply chain level, are also introduced in this chapter and given a precise definition. A basic model to represent supply chain dynamics is constructed on a step-by-step basis in Chap. 4, where the main hypotheses and limitations of the basic model are discussed. From this model, the quality of the information acquired by each node and the information transmitted to the rest of the nodes in the supply chain emerges as a key factor influencing its dynamic performance. In Chap. 5, this aspect is modelled and discussed in great detail, and information-related aspects are modelled and presented. In Chap. 6, the basic models presented in Chap. 4 are enriched by removing some hypotheses and including additional constraints in order to contemplate a wider range of scenarios observed in practice, including variable lead times or capacity constraints. Chapter 7 is devoted to modelling the dynamics of the so-called closed-loop supply chains, where there is a reverse flow of materials and information. In Chap. 8, the bullwhip effect related to the existence of non-linear supply chain structures is addressed. Finally, further topics within the area are discussed in Chap. 9.

Seville, Spain
May 2021

Jose M. Framinan

Acknowledgements

At the time I write these lines, I try to remember how the world was when I started drafting the first pages by the winter of 2019, and I have just realised that the book has been written practically during a lockdown or under heavy mobility restrictions. During this time, I (as almost everyone) had to face a completely new situation that forced us to change our way to work and to live, a fact hardly worth mentioning when compared to the impact on many other people around the world. Let us hope that, very soon, we can look back at these times with the same perplexity as now I look back at the winter of 2019.

Writing this book in these unpleasant times has been made easier by a number of people, whom I a very grateful. First, I would like to thank the editorial staff at Springer, particularly Anthony Doyle and Jayanthi Krishnamoorthi for being extremely responsive and patient with me as I failed to meet the successive deadlines. I am also thankful to Salvatore Cannella and Roberto Dominguez, both former Ph.D. students of mine, for working with me on these topics for a long time, now with the addition of Borja Ponte. The rest of my research group also should be thanked for many reasons, not being the less important to cope with me as an absent boss. And above all, I have received love and support from my family. Bringing work home has not changed it as the place where I always want to stay.

Contents

Acronyms

3PL	Third Party Logistics
ADI	Advance Demand Information
AI	Artificial Intelligence
AM	Additive Manufacturing
APICS	American Production and Inventory Control Society
ATO	Assemble-To-Order
BTO	Build-To-Order
CLSC	Closed-Loop Supply Chain
CPS	Cyber-Physical System
CRP	Continuous Replenishment Program
CTO	Configure-To-Order
CT-TP	Cycle Time-Throughput
DTO	Design-To-Order
ECR	Efficient Consumer Response
EDLP	Every Day Lowest Price
ETO	Engineer-To-Order
FTL	Full Truck Load
HSC	Hybrid Supply Chain
ICT	Information and Communication Technologies
IoT	Internet of Things
IRI	Inventory Record Inaccuracy
IS	Information System
MAE	Mean Absolute Error
MAPE	Mean Absolute Percentage Error
ME	Mean Error
MMSE	Minimal Mean Squared Error
MPE	Mean Percentage Error
MRO	Maintenance, Repair and Overhauling/Maintenance, Repair and Operations
MSE	Mean Squared Error
MTO	Make-To-Order
MTS	Make-To-Stock

ORV	Order Rate Variance
ORVR	Order Rate Variance Ratio
OUT	Order-Up-To
POS	Point Of Sale
PTM	Priority To Manufacture
PTR	Priority To Remanufacture
RFID	Radio Frequency Identification
RMSE	Root of Mean Squared Error
RQU	Returns Quality Uncertainty
RTU	Returns Times Uncertainty
RVU	Returns Volume Uncertainty
SC	Supply Chain
SCC	Supply Chain Council
SCOR	Supply Chain Operations Reference
SKU	Stock Keeping Unit
VMI	Vendor Managed Inventory

Chapter 1
Introduction to Supply Chains

1.1 Introduction

In this chapter, we discuss the evolution of the business and manufacturing scenarios that have led to the prevalence of supply chains as a form of competition. A basic definition of a supply chain as a set of interrelated companies that produce and distribute the goods and services to a final customer is given, along with some examples of real-life supply chains in operation. Next, the need of managing the whole supply chain rather than the individual companies is presented, and some pioneering experiences are briefly commented. Then, it is discussed that, along with the advantages in terms of efficiency brought by the supply chain paradigm, early experiences also detected undesirable effects regarding inventory and demand amplification (i.e. the bullwhip effect) induced by the dynamics of the supply chain. The relevance and impact of this effect on supply chain performance are described. As a result, the need to model the dynamics of the supply chain to fully understand the causes of these detrimental effects and to improve supply chain performance is highlighted.

As a summary, in this chapter, we

- Present the main characteristics and definitions of a supply chain, as well as the factors that have led to the prevalence of this form of business (Sect. 1.2).
- Introduce the bullwhip effect as the main factor driving the dynamics of the SC, giving a hint on its causes and barriers to overcome it (Sect. 1.3).
- Discuss the need of models to study SC dynamics (Sect. 1.4).

1.2 The Rise of Supply Chains

Regardless of the economic sector, supply chains or supply networks constitute nowadays the preferred form of manufacturing and distribution of products and services. Even if a more formal definition of a Supply Chain (SC) would be given later, this

J. M. Framinan, *Modelling Supply Chain Dynamics*,
https://doi.org/10.1007/978-3-030-79189-6_1

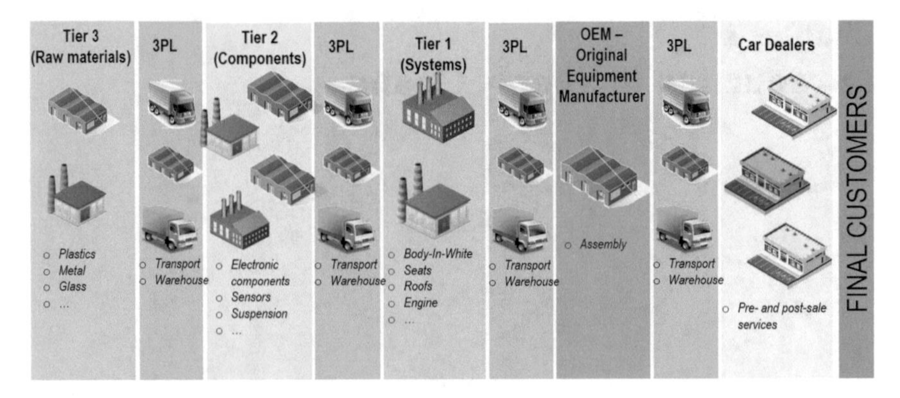

Fig. 1.1 Typical elements in an automotive supply chain

type of production-distribution is characterised by the decentralisation of the different activities required to manufacture a product and to deliver it to the final customer. Figure 1.1 shows a typical SC in the automotive sector. As it can be seen in the figure, different companies play a highly specialised role in the manufacturing and distribution of a car.

This organisation of the production process is far from being exclusive of certain sectors, but it is rather commonplace. From an abstract viewpoint, it consists of splitting the production process into a sequence of assembly and sub-assembly operations taking place in different facilities (that may pertain or not to the same company) whose flow of materials is handled by parties offering logistic (mainly distribution and warehousing) services. The range of services offered by these logistic partners serves to define their level (see Fig. 1.1), and the proximity of the manufacturing company to the final assembly defines the levels (*tiers*) in the SC.

From a business viewpoint, the SC means taking a whole product and sharing across different companies the responsibilities for its manufacturing and distribution. In some cases, even the company responsible for the product externalises all tasks related to the manufacturing and logistics of the product, focussing exclusively on those with higher added-value. This can be seen in Fig. 1.2, which depicts a simplified version of the different companies collaborating in the SC of one of the models of Apple's iPhone. As it can be seen, Apple only retains the R & D and design activities, and some parts of the distribution and sales. Indeed, some of Apple's products—those sold via the online sales channel—do not get to visit a single Apple facility.

1.2.1 A Definition of Supply Chain

There are many definitions of the supply chain, each one depending on the specific aspect that is to be highlighted. For the purposes of the book, we will stick to a rather simple definition of the supply chain as a set of interrelated companies that

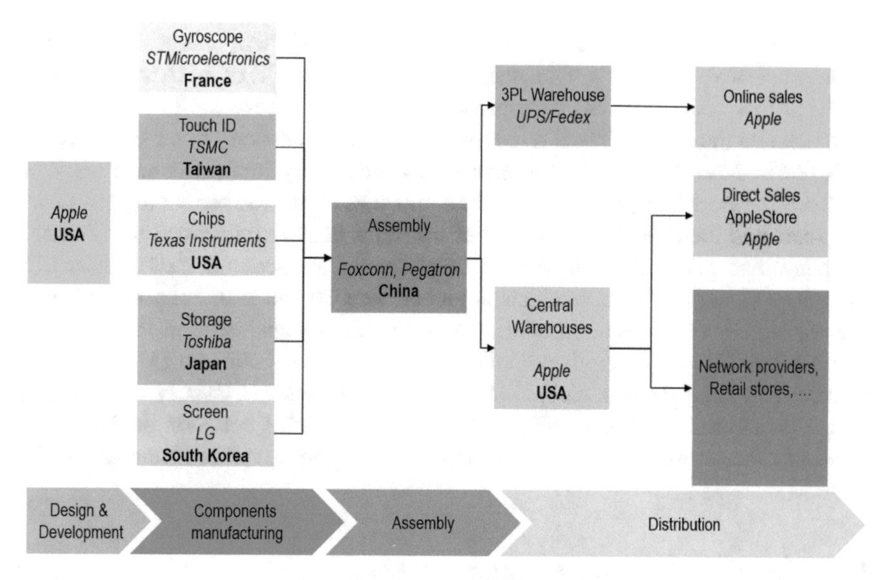

Fig. 1.2 Typical elements in Apple's iPhone SC

produce and distribute the goods and services to a final customer. The interrelation of the companies refers to the fact that, even if these companies may be completely separate legal entities, their different products and processes create value for a final (common) customer. Note that, with this definition, we are adopting an end-to-end perspective of the flows of products and information, not only from the source of raw materials to the delivery of the final product to the customer, but also including in some cases the after-sales and the return of the products at the end of their life cycle. Particularly, this latter flow of materials is becoming increasingly important, and indeed, it will be discussed in detail in Chap. 7.

From an organisational viewpoint, the SC is a hybrid between a hierarchical organisation and a pure market-based relationship among companies. Arranging the production and distribution of goods in a hierarchical organisation typically leads to a more coordinated flow of material and information among the different facilities in the organisation as there are common goals that usually translate into a global plan which is adopted by all parties. However, it is difficult to balance the capacity of the different facilities in such organisations, and the lack of competition with external companies may lead to poor overall performance. In contrast, the advantages provided by a market-based relationship among companies (typically regulated by short-term supplier-customer contracts) in terms of efficiency are counterbalanced by the high instability and complexity of this type of organisation, since it is difficult to establish long-term relationships among the members. Furthermore, as we will discuss, some methods to avoid undesirable inventory costs are based on the accurate and timely exchange of information among companies, for which a high level of trust is required. Clearly, a relationship among companies purely based on the market does not foster the establishment of such trust. The SC then emerges as a hybrid organisation that, if properly managed, could reap the benefits of these two extremes.

Using this concept of SC as a broad middle ground between the hierarchical approach and the market-based relationship raises the question of how close are the SC to one of these extremes. In this regard, there are usually great differences between the so-called *focal* SCs (where one of the companies is able to exert more power over the others and therefore the operation of the SC is heavily influenced by this focal company) and other SCs where such focal companies does not exist. Focal SCs are indeed closer to the hierarchical approach whereas the market-based approach is more similar otherwise. An additional aspect to be considered in this regard is whether the SC is *intraorganisational* (i.e. it is composed of companies belonging to the same industrial group or cluster) or it is *interorganisational*. Clearly, in the former case, it is easier to develop a common plan or at least some common guidelines. In the latter case, coordination is usually achieved using more subtle mechanisms.

1.2.2 Factors Facilitating Supply Chains

As we have seen, SCs are complex organisations regardless of their type. Such an intricate network of relationships among companies has not appeared overnight. Indeed, a number of business trends have been (and are still) boosting SCs. Among these, we can mention the following:

- **Globalisation**. Although this may not be a trend in the future, according to the World Trade Organisation, the number of trade agreements among countries and regions has increased steadily over the years. Although the extent and nature of these trade agreements may be very different from one agreement to another, the overall effect is the reduction of the import taxes and, consequently, an increase of the competition faced by companies, as in most cases, they no longer compete solely with the local/regional players in the market, but with companies across the world.
- **Market maturity**. For many sectors, the market can be considered mature, meaning that the demand for the products in these sectors has reached a state of equilibrium and there is no significant growth. In this absence of substantial growth, competition becomes more acute to increase their market share. Although it is not the only strategy to do so, one option for the companies is to become extremely efficient in their operations, thus outsourcing all activities where the company does not excel. Typically, this may involve not only subcontracting services or supporting activities, but also some parts of their production process.
- **Increase in product/service complexity**. The complexity of the products/services provided to the final customer has increased in many cases and, as a consequence, providing an excellent tuple product-service requires a wide range of skills and competences that may not be found in a single company, particularly if this is coupled with the intense competition experienced due to the globalisation and market maturity discussed above.

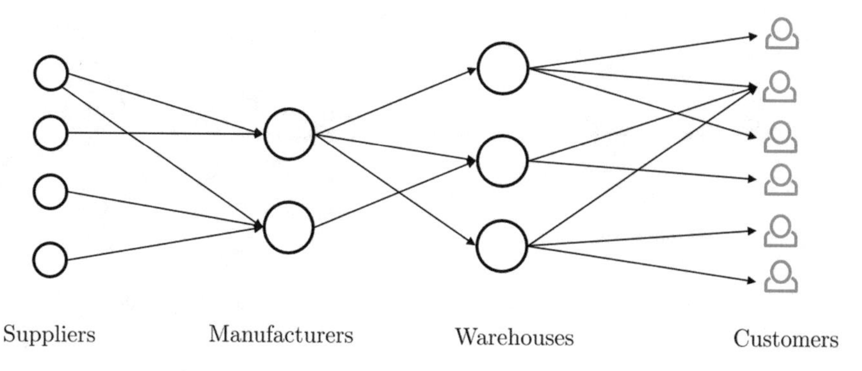

Fig. 1.3 Depiction of a SC model

- **ICT (Information & Communication Technologies) growth**. Naturally, the effective coordination of activities across the SC requires a reliable ICT infrastructure, as otherwise the synchronisation of the material and information flows among companies becomes a nightmare. Formerly an entry barrier for many companies, such infrastructure has become nowadays ubiquitous and inexpensive. The standardisation of communication protocols around the Internet has also led to a substantial reduction of the otherwise extremely expensive interoperability costs.

The juxtaposition of these trends, among others, has made the business scenario highly competitive and has led the customers with high power. The almost universally adopted solution for the companies has been to focus on these processes of the product for which the company was able to excel (*core competences*) and to outsource to other firms the manufacturing of these parts of the product/service that could be better offered by another specialised company.

1.2.3 Main Elements of a Supply Chain

For modelling purposes, SCs are usually represented as a network depicting each stakeholder taking part in the process as well as their (usual) flow of materials, such as shown in Fig. 1.3. Each stakeholder is referred to as a *node*, while the roles of the stakeholders—facilities, warehouses, distribution centres, etc.—are usually grouped vertically and are collectively denoted as *echelons* or levels.[1] In this representation, products flow from the left side to the right side, so it is usual to denote as *downstream* (*upstream*) the echelons in the right (left) size of a node.

[1] Another term mentioned before—*tier*—is much more common in industry, whereas the term echelons is prevalent in the academic literature, and it may have a slightly different meaning. In this book, both terms would be used indistinguishably.

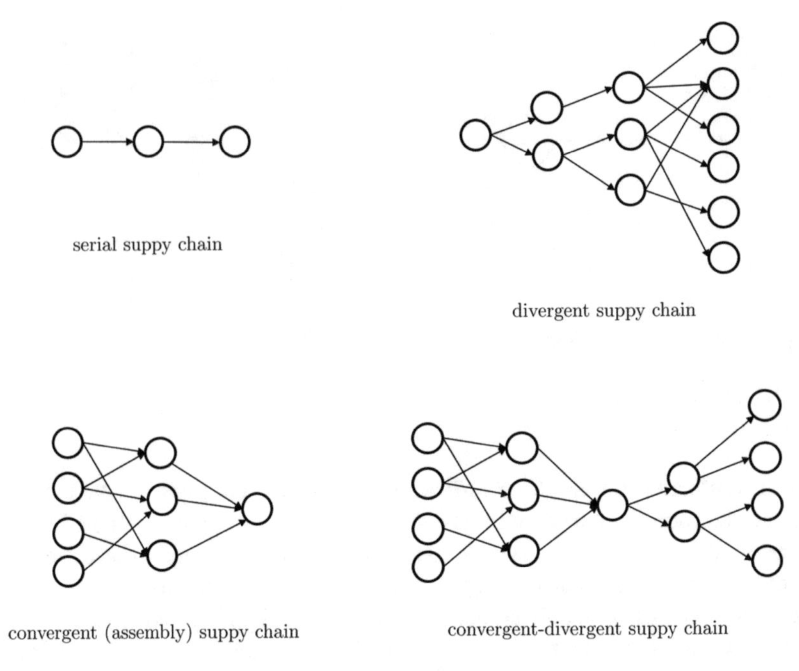

Fig. 1.4 Different structures of a SC

As it can be seen from the figure, the name supply chain is a simplification, as it implies a linear relationship between the nodes whereas it would be more appropriate to refer to a *supply network* to stress their, in general, non-linear relationship. However, the name supply chain is much more common and we will stick to it in this book. Indeed, to refer to a linear supply chain, we will use the term *serial supply chain* or *serial system*, in order to differentiate it from other usual supply chain structures, such as divergent, convergent or convergent-divergent. Figure 1.4 shows some examples.

When the models are developed in the next chapters of the book, we adopt the convention of numbering the echelons according to their proximity to the final customer. Therefore, echelon 1 is formed by the nodes directly interacting with the final customer, whereas echelon 2 is formed by the nodes interacting with nodes in echelon 2.

1.3 New Problems, New Opportunities: The Bullwhip Effect

Although the formation of SCs can be seen as the *natural* adaptation of the companies to a high-competition business environment, this has not come without troubles.

Clearly, it is more difficult to synchronise the flow of materials and information in a single company, than for different companies across the SC.

Finally, a shift in the management paradigm has also had to be carried out. For a company operating in a SC, the traditional wisdom depicting a customer as king (and, consequently, the company's provided as the opposite) has to be modified. In a SC, there is only one *real king*, as only the final customer is bringing in external money to the SC while the rest of the nodes simply allocate this money among them. Another way to put it is to state that the competition is no longer among companies, but among supply chains. In a time where the final customer has the power to choose, there is really no difference if the cause of his/her dissatisfaction with the product that he/she has just acquired is due to defective component processing, poor assembly or late delivery: the result is always the loss of competitiveness of the final product, and consequently, the reduction of the external money entering the SC. With this in mind, it is clear that the focus should be put on the satisfaction of the final customer and, as a consequence, local (intra-company) actions aimed at fulfilling internal objectives may not be deemed as positive for the SC if these do not result in improved performance of the overall SC.

With this shift in the management paradigm, suitable metrics to measure the efficiency of the operations of the entire SC have to be put in place. As a consequence, the effect of local (i.e. intra-company) actions on the overall SC has to be analysed. This aspect is extremely important in SCs since they consistently suffer a well-known phenomenon that affects the members of the SC in an asymmetric manner. This phenomenon is known as the *bullwhip effect* and will be discussed in the next subsection.

1.3.1 The Bullwhip Effect

The so-called bullwhip effect is perhaps the most specific (and undesirable) effect deriving from the SCs as a way to organise the production and distribution of products. Although the stories regarding the history of the bullwhip effect may diverge, it seems that one of the best-documented earlier recollections of this effect was detected by Procter & Gamble executives while studying the demand trends of baby diapers. The number of babies born does not exhibit a clear seasonality pattern across the year,[2] so it seems reasonable to expect a more or less stable demand of diapers from the customers.

This behaviour was effectively observed by looking at the inventory in the supermarkets. However, it turned out that the orders placed by the supermarket to the distribution centres were more variable (i.e. higher peaks and deeper valleys) than the demand of the consumer. This pattern was sharper when looking at the orders placed by the distribution centres to the manufacturer and was increasingly acute

[2] Indeed there are some studies arguing that there is some seasonal variation in the number of births, but nevertheless such variation is rather small.

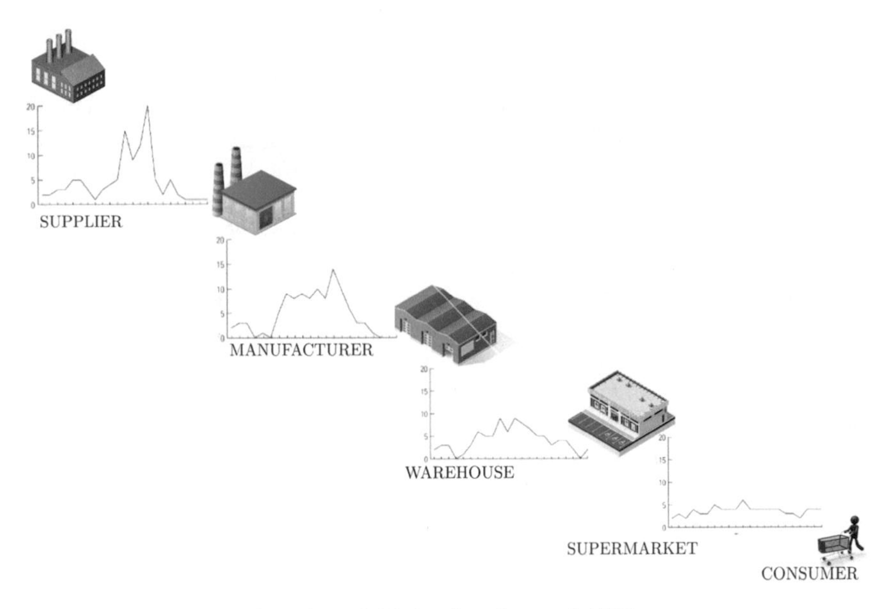

Fig. 1.5 Demand patterns in a Diaper SC (taken from Lee et al. 1997)

when moving to the supplier of the manufacturer. Figure 1.5 illustrate the behaviour of the different patterns in the demand. As it can be seen, these patterns resemble a whip, hence Procter & Gamble executives coined the term *bullwhip effect*.

Although it is clear that, on average, the demand was the same across all nodes in the SC, this increase in order variability has acute economical implications. For most products, the cost of not being able to serve in time a unit required by the customer is usually much higher than the unit inventory costs, inventory is built to protect the system against the peaks in demand. Roughly speaking, that means that the higher the variability in demand, the most costly it is for the company to face the same average demand. Furthermore, these costs may strip out the potential gains that might be obtained from other initiatives. For instance, Barilla SpA, a major Italian pasta producer, offered special discounts to customers ordering full truckload quantities. However, such a deal created customer order patterns that were so erratic and unpredictable that they ate the benefits created by full truckload transportation.

This observed increase in order variability has been documented across different sectors,[3] also with disperse magnitude of amplification: for instance, in the food sector, it has been found that the supplier orders two tiers further upstream in the SC varied 10 times more than those in the points of sales, whereas in the automotive sector, the ratio of the variance between incoming orders and the orders placed to the suppliers was 1:2. In general, the bullwhip effect has been found almost in any economic sector, and some studies calculate that around 2/3 of companies are affected by the bullwhip effect, which provokes around 1/3 of inventory extra costs.

[3] Indeed, it has not only be observed in manufacturing industries, but also in services.

Surely, one might expect that a widespread effect has a different number of causes, since these companies have different operational practices. Traditionally, the causes of the bullwhip effect have been classified either as operational or behavioural.

Behavioural causes were the first to be explored as a cause of the bullwhip effect. Roughly speaking, this stream of causes blames the managers in the companies for what we might call a *non-rational* behaviour. Some examples of this behaviour—discussed with more detail in Sect. 3.3 of Chap. 3—include the overstatement of anecdotal data, the over- or under-reaction to changes in demand, risk aversion or cognitive limitations due to the inherent complexity of the SCs.

The behavioural causes of the bullwhip effect were exemplified in the 80s using the so-called *Beer Distribution Game*, initially as an experiment with the class students to gauge the effect of the aforementioned non-rational behaviour in the supply chain. In this game, each student or group of students is in charge of a facility (factory, distributor, wholesaler and retailer) within a SC that must deliver the product to the final customer. Even if the demand from the final customer was relatively stable, the non-rational behaviour of the students in charge of the facilities use lead to an extremely poor performance of the overall SC (typical operational costs in a game would be 10 times the optimal costs). Furthermore, the variance of the orders would amplify as one moves upstream in the SC.

From the behavioural perspective, the solution to the bullwhip effect might seem relatively straightforward: as the root cause is the non-rational behaviour of the managers in the supply chain, let us teach the managers to behave rationally.[4] However, a series of papers by Hau Lee in the 90s showed that there are also *operational* causes of the bullwhip effect, i.e. the bullwhip effect may appear even if all managers across the supply chain behave rationally, even if they make optimal or near-optimal decisions. More specifically, he identified four major operational causes. These causes will be discussed in detail later in the book (first intuitively in Sect. 3.4 in Chap. 3, and later more extensively throughout the book), but we can say that these are generally related to the need to face an uncertain demand (and therefore to make replenishment decisions under uncertainty), coupled with (again uncertain) expected variations in the price or the regularity of the replenishments. Perhaps the most striking feature of this type of cause is that it is precisely the *rational* (or optimal) behaviour of the stakeholders in the SC which is one of the reasons that these causes exacerbate the bullwhip effect.

Both behavioural and operational causes might co-exist, thus exacerbating the demand variability pattern and their negative effects on the profitability of the SC. Furthermore, as we will discuss in the next section, even if the causes of the bullwhip effect (or at least some of them) have been identified and some potential solutions have been devised, there are a number of barriers making them difficult to implement.

[4] However, there is ample evidence that, although experience may mitigate the behavioural causes of the bullwhip effect, it does not serve to remove it (check Chap. 3 in this regard).

1.3.2 Barriers to Overcome the Bullwhip Effect

There are a number of barriers to overcome the bullwhip effect, including the following:

- First of all, its effects are not equally perceived by all nodes in the SC. Since the orders placed upstream by an echelon in the SC constitute the demand seen by the upstream echelon, it is clear that the manufacturer or suppliers suffer a higher increase in the costs, while the downstream echelons may not be affected at all. Furthermore, some of the solutions proposed to tackle the bullwhip effect may result in a decrease in the service level at the retailer. Therefore, a co-operation scheme beneficial for all nodes is required: for instance, the retailer may be benefited by, e.g. lower purchasing costs from a supplier who has seen its inventory costs reduced. To set up these schemes, a global vision of the SC overcoming the traditional, intra-company wide operations is required.
- Secondly, not all SCs suffer equally the bullwhip effect, so it has to be measured and quantified to know its extent and to balance the potential savings against the costs of the measures to reduce its damage. Several measures of the bullwhip effect will be discussed in Chap. 4.
- As we will discuss later in the book, many techniques to reduce or alleviate the consequences of the bullwhip effect are based on sharing information (such as, e.g. demand data or inventory levels) among the members of the SC, which clearly requires a substantial amount of trust in order to share such sensitive information. Again, challenging the traditional view of not sharing sensitive company data requires a substantial managerial change that is not always possible to achieve due to the companies' inertia.

1.4 Modelling Supply Chain Dynamics

We have already discussed that the term SC encompasses a wide variety of business models, with rather different coordination and operation mechanisms. In view of such variety, it does not seem sensible to rely purely on empirical research or field studies to address the challenges in SC management, as it is extremely difficult to differentiate sector- or product-specific effects from structural ones. An alternative is to study SC dynamics using models, which can be employed to validate hypotheses observed in empirical studies or to explore SC management policies that it would be impossible to test otherwise.

Therefore, the role of models in understanding supply chain dynamics is very important. As it has been discussed before, there are a number of factors influencing the dynamic behaviour of a SC. This number increases as one moves from very simple 'toy' SC (such as the one presented in the beer game) to more realistic SCs where additional factors (such as the lack of timely, accurate information or the return flow of goods) have to be considered. Models are key to understand the role played by these factors.

However, developing a suitable SC model is not an easy task. First of all, SC is formed by the interaction of material, information and money flowing from different entities, each one with its own business and decision processes. Therefore, even an extremely simple SC model should take into account diverse business functions such as demand forecasting, inventory management or replenishment policy (among others) *for each node* of the SC, as well as to properly capture the interaction among these nodes. At the same time, not all aspects of a real-life SC can be embedded in the model, since extremely complex models may not be fit to extract general guidelines for managers, as the conclusions may be heavily dependent on the specific parameters of the model. Therefore, a trade-off between these two aspects may have to be found.

In this book, our preferred choice would be to develop relatively simple analytical models to understand the main factors and variables affecting SC dynamics. However, we will use also simulation models whenever the phenomenon requires so. We resume this discussion on the merits of both approaches in Sect. 4.2 of Chap. 4.

1.5 Summary

As we have discussed in this chapter, SCs constitute the prevalent business paradigm for many sectors and products, and it has been fostered by a number of factors and market trends such a globalisation, market maturity or the growth in ICT. It consists of a set of interrelated companies collaborating in the production and distribution of the goods and services to a final customer, and usually results in a rather complex flow of materials, information and money across the companies belonging to the SC. Although the benefits of this business paradigm are clear (ranging from a higher efficiency than hierarchical organisations to a more focussed and stable arrangement than a pure market-based relationship), SCs are not problem-free. Perhaps the more ubiquitous and persistent problem is the bullwhip effect, which designates the amplification of the variability of orders and inventory across the SC. The bullwhip effect causes important costs in terms of extra inventory and hinders the service level of the members of the SC. The causes of the bullwhip effect range from the so-called behavioural causes—i.e. related to the non-rational (sub-optimal) behaviour of the Decision Makers in the SC—to structural causes—i.e. inherent to the structure of the SC even if the decisions are taken in an optimal manner. The complexity of this phenomenon and the variety of causes make necessary its in-depth analysis using models that encompass the different activities of the nodes in the SC.

1.6 Further Readings

There are many books and texts discussing the definition of SC and its evolution over time. Among them, it is worth highlighting [1] or [2]. Several concepts used in this chapter can be found in these books.

Historical recollections, together with analysis and future research lines on the bullwhip effect, can be found in [3] or [4]. The case of Barilla SpA presented in Sect. 1.3.1 is described in [5, 6]. The data regarding the variability of the amplification in the automotive sector is from [7], whereas the ratio 1:10 in the food sector is documented in [8]. Documented cases of the bullwhip effects in the service sector are presented in [9, 10]. Sectorial studies on the impact of bullwhip effect include [11] for the car industry, [12] for the machine tool sector, the aggregated manufacturing sector [13, 14] for the computer and semi-conductor industry or [15] for the retail sector. Intra-industry bullwhip (i.e. determining which business function within a company is the source of bullwhip) is analysed in [16, 17]. Other empirical evidence of the manifestation of the bullwhip effect can be found, e.g. in [18] or in [19]. Note however that, even if the bullwhip effect clearly increases the operating costs, its translation into a decline of the overall performance of the firm is not direct [20].

It is customary to present the works by Forrester [21, 22] as the first academic descriptions of the bullwhip effect, although in the same year, Burbridge [23] presented a methodology for controlling production and inventory which was linked to the problem of 'demand amplification', as it was defined by Forrester. Other authors (e.g. [24]) even point out to Mitchell who, three decades before, mentioned the 'deception and illusion' due to 'over-ordering' as an obstacle to see the picture of the true demand [25].

The seminal reference for the beer game is [26], although other authors have extended this game or produced computer-based versions (see, e.g. [27, 28] or [29]). Some studies on the behavioural causes of the bullwhip effect are due to [26, 30]. The seminal papers on the operational causes of the bullwhip effect are [31, 32]. The organisational barriers to collaboration among industries in the SC are studied in [33].

References

1. Stadtler, H., Kilger, C.: Supply chain management and advanced planning (Fourth edition): Concepts, models, software, and case studies (2008)
2. Snyder, L., Shen, Z.M.: Fundamentals of Supply Chain Theory, 2nd edn. (2019)
3. Geary, S., Disney, S., Towill, D.: On bullwhip in supply chains - historical review, present practice and expected future impact. Int. J. Prod. Econ. **101**(1 SPEC. ISS.), 2–18 (2006)
4. Wang, X., Disney, S.: The bullwhip effect: progress, trends and directions. Eur. J. Oper. Res. **250**(3), 691–701 (2016)
5. Hammond, J.: Barilla spa. In: Harvard Business School Case 6-694-046. Boston (1994)
6. Disney, S., Lambrecht, M.: On replenishment rules, forecasting, and the bullwhip effect in supply chains. Found. Trends Technol. Inf. Oper. Manag. **2**(1), 1–80 (2007)

7. Naim, M., Disney, S., Evans, G.: Minimum reasonable inventory and the bullwhip effect in an automotive enterprise; a "foresight vehicle" demonstrator. SAE Technical Papers (2002)
8. Jones, D., Simons, D.: Future directions for the supply side of ecr. ECR in the Third Millennium - Academic Perspectives on the Future of Consumer Goods Industry pp. 34–40 (2000)
9. Akkermans, H., Vos, B.: Amplification in service supply chains: an exploratory case study from the telecom industry. Prod. Oper. Manag. **12**(2), 204–223 (2003)
10. Anderson, E., Jr., Morrice, D.: A simulation game for teaching service-oriented supply chain management: does information sharing help managers with service capacity decisions? Prod. Oper. Manag. **9**(1), 40–55 (2000)
11. Blanchard, O.: The production and inventory behavior of the american automobile industry. J. Pol. Econ. **91**(3), 365–400 (1983)
12. Anderson E.G., J., Fine, C., Parker, G.: Upstream volatility in the supply chain: the machine tool industry as a case study. Prod. Oper. Manag. **9**(3), 239–261 (2000)
13. Dooley, K., Yan, T., Mohan, S., Gopalakrishnan, M.: Inventory management and the bullwhip effect during the 2007–2009 recession: evidence from the manufacturing sector. J. Supply Chain Manag. **46**(1), 12–18 (2010)
14. Terwiesch, C., Ren, Z., Ho, T., Cohen, M.: An empirical analysis of forecast sharing in the semiconductor equipment supply chain. Manag. Sci. **51**(2), 208–220 (2005)
15. Gill, P., Abend, J.: Wal-mart: the supply chain heavyweight champ. Supply Chain Manag. Rev. **1**(1), 8–16 (1997)
16. Tesfay, Y.: Modeling the causes of the bullwhip effect and its implications on the theory of organizational coordination. Supply Chain Forum **16**(2), 30–46 (2015)
17. Jin, M., DeHoratius, N., Schmidt, G.: In search of intra-industry bullwhips. Int. J. Prod. Econ. **191**, 51–65 (2017)
18. Isaksson, O., Seifert, R.: Quantifying the bullwhip effect using two-echelon data: a cross-industry empirical investigation. Int. J. Prod. Econ. **171**, 311–320 (2016)
19. Wu, D., Teng, J., Ivanov, S., Anyu, J.: Empirical assessment of bullwhip effect in supply networks. Int. J. Inf. Syst. Supply Chain Manag. **14**(2), 69–87 (2021)
20. Mackelprang, A., Malhotra, M.: The impact of bullwhip on supply chains: performance pathways, control mechanisms, and managerial levers. J. Oper. Manag. **36**, 15–32 (2015)
21. Forrester, J.: Industrial dynamics: a major breakthrough for decision makers. Harv. Bus. Rev. **36**(4), 37–66 (1958)
22. Forrester, J.: Industrial Dynamics (1961)
23. Burbidge, J.: The new approach to production. Prod. Eng. **40**(12), 769–784 (1961)
24. Chatfield, D., Pritchard, A.: Returns and the bullwhip effect. Trans. Res. Part E: Log. Trans. Rev. **49**(1), 159–175 (2013)
25. Mitchell, T.: Competitive illusion as a cause of business cycles. Quart. J. Econ. **38**(4), 631–652 (1924)
26. Sterman, J.D.: Modeling managerial behavior: misperceptions of feedback in a dynamic decision making experiment. Manag. Sci. **35**(3), 321–339 (1989)
27. Van Ackere, A., Larsen, E., Morecroft, J.: Systems thinking and business process redesign: an application to the beer game. Eur. Manag. J. **11**(4), 412–423 (1993)
28. Kaminsky, P., Simchi-Levi, D.: A new computerized beer game: a tool for teaching the value of integrated supply chain management. Supply Chain Technol. Manag. **1**(1), 216–225 (1998)
29. Lambrecht, M., Dejonckheere, J.: A bullwhip effect explorer. Research Report 9910 (1999)
30. Croson, R., Donohue, K.: Experimental economics and supply-chain management. Interfaces **32**(5), 74–82 (2002)
31. Lee, H., Padmanabhan, V., Whang, S.: Information distortion in a supply chain: the bullwhip effect. Manag. Sci. **43**(4), 546–558 (1997)
32. Lee, H., Padmanabhan, V., Whang, S.: The bullwhip effect in supply chains. Sloan Manag. Rev. **38**(3), 93–102 (1997)
33. Fawcett, S., McCarter, M., Fawcett, A., Webb, G., Magnan, G.: Why supply chain collaboration fails: the socio-structural view of resistance to relational strategies. Surg. Endosc. Other Interv. Tech. **20**(6), 648–663 (2015)

Chapter 2
Supply Chain Management

2.1 Introduction

As we have seen in Chap. 1, supply chain dynamics originates from the interaction of different business functions both taking place within the company and also in the company's relationship with their customers and providers. Understanding these functions is key to develop accurate models that help us to gain insights into this field. Therefore, in order to make it possible to read the book with minimal background, the aim of this chapter is to equip the reader with the background on operations management required to understand the supply chain models to be developed in the next chapters. To do so, we use the SCOR model to identify the main activities taking place in the supply chain (i.e. source, make, deliver and return) and discuss the topics involving each one of these main activities. As we will see, 'deliver' activities are discussed with a focus on demand forecasting, so the most usual methods for demand forecasting and modelling are briefly introduced, along with the typical measures of the forecast error. The main concepts from inventory management employed in the models in the rest of the book (order-up-to, safety stock determination, smoothing, etc.) are discussed as part of the 'source' activities.

More specifically, in this chapter, we:

- Describe the main activities that take place in the SC by means of the SCOR model, a widely used reference model to describe SC operations (Sect. 2.2).
- Present the most used methods for demand modelling and forecast, and develop some useful formulae that will be used in subsequent chapters (Sect. 2.3).
- Discuss inventory management techniques, with an emphasis on stock-base replenishment policies (Sect. 2.4).

J. M. Framinan, *Modelling Supply Chain Dynamics*,
https://doi.org/10.1007/978-3-030-79189-6_2

2.2 Main Activities in SCM: The SCOR Model

From a managerial, high-level viewpoint, we can think of a SC as a sequence of interlinked activities required to satisfy the final customer demand, regardless of the specific products or services constituting this demand. At a high level, these activities can be grouped within one (or more) of the following sets of related activities (that, for the moment, we will denote as *management processes*):

- Source: Source processes are in charge of procuring goods and services to meet (actual or planned) demand.
- Make: Make processes involve the transformation of a product from an initial (not available for sales) state to a final state (finished good) where it can meet the demand.
- Deliver: Deliver process involves the provision of the finished good to satisfy the demand.
- Return: It includes the processes associated with returning products that, for some reason, do not meet the demand.
- Plan: It includes the processes associated with balancing demand and supply.

These activities form the basis of the so-called SCOR (Supply Chain Operations Reference) model, a process reference model developed by the Supply Chain Council (now a part of APICS) as the standard for supply chain management that is shown in Fig. 2.1. The Supply Chain Council was organised in 1996 and it is a not-for-profit trade association in which the majority of the members are companies representing a broad cross-section of industries, including manufacturers, distributors and retailers. The first version of the SCOR model was introduced in 1996 and since then it has become a de facto professional standard for the description of SCs, and it is widely used by companies and organisations around the world to analyse and improve their operations. In the latest version, a sixth activity named *Enable* has been introduced to encompass the processes related to the management of the SC, see Fig. 2.2, where also some minor semantic changes can be observed.

The SCOR model defines three levels and describes a fourth level related to implementation issues, although this level is company-specific and therefore outside the scope of a reference model. Level 1 (also denoted as scope level) is the highest one

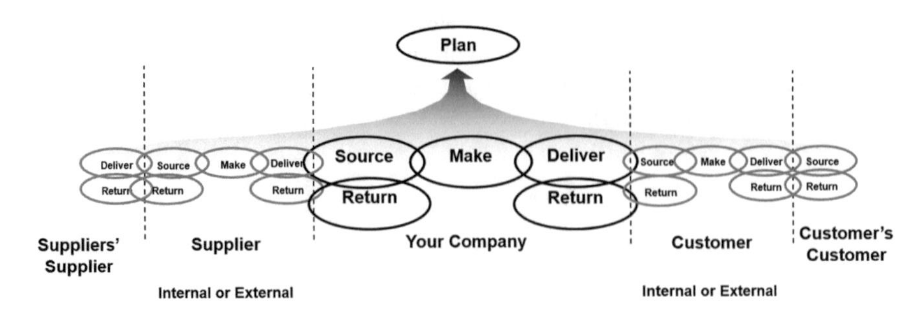

Fig. 2.1 Original SCOR model [Supply Chain Council 2000]

Fig. 2.2 SCOR model from revision 11 on [Supply Chain Council 2012]

and serves to identify the process types or *business processes* in the SC.[1] Consequently, it consists of the six activities mentioned before. These six processes can be carried out by a company in different ways depending on what it is called the *configuration* of the process. More specifically, for the Source, Make and Deliver process, this configuration depends on the manufacturing strategy adopted for the products/services handled by the node. In this regard, three manufacturing strategies (or *environments* in the SCOR jargon) are identified in the SCOR model:

- Make-To-Stock (MTS). Products in a MTS environment are expected to be shipped from finished goods or *off the shelf* and thus completed prior to the receipt of a customer order. As a consequence, they are generally produced according to a sales forecast. As a rule, a final product does not have a customer reference or customer order detail.
- Make-To-Order (MTO). The manufacturing of products in a MTO environment is initiated only in response to customer order. Therefore, the product—at any stage of its processing—is traceable to a specific customer. In this environment, it is assumed that the customer expects the product to be delivered in a lead time which is equal or higher to the cycle time of the product. In the SCOR model, the MTO environment also encompasses similar or related manufacturing strategies such as Build-To-Order (BTO), Assemble-To-Order (ATO), Configure-To-Order (CTO) or postponement.
- Engineer-To-Order (ETO). The development, design and manufacturing processes to produce products or services are based on the requirements of a specific customer. It is generally employed for highly customised products and, in general, it is required that work instructions for the different manufacturing operations are defined or refined, and material routing instructions may need to be added or modified. In the SCOR model, an alternative name for ETO is Design-To-Order (DTO).

[1] Note however that the SCOR model only describes *operations* in the SC and it does not attempt to describe other business processes or activities in the companies such as sales & marketing, product development, research and development or post-delivery customer support.

Table 2.1 Example of levels in the SCOR model

Level	Code	Description
...	...	...
1	sM	Make
2	sM1	Make-To-Stock
3	sM1.1	Schedule production activities
3	sM1.2	Issue material
3	sM1.3	Produce and test
3	sM1.4	Package
3	sM1.5	Stage product
3	sM1.6	Release product to deliver
3	sM1.7	Waste disposal
2	sM2	Make-To-Order
...	...	...
2	sM3	Engineer-To-Order
...	...	...

Clearly, the manner in which activities such as procuring the raw materials or delivering the final products to the customer are carried out is heavily depending on the manufacturing strategy adopted by the node. At the same time, usually, a company would not adopt all of these manufacturing strategies so, while all companies are expected to have a Make process at level 1 (or not to have it), how this Make process is *configured* or implemented in a node is company-specific. For this reason, level 2 is also denoted as Configuration level and it consists of a couple of dozen (depending on the version of the SCOR model) sub-processes which are *children* of the corresponding level-1 process. As an example, the level-2 processes children of the Make process are Make-To-Stock, Make-To-Order and Engineer-To-Order.

In the case of the Plan process, the configuration refers to planning the whole SC as well as the Source, Make, Deliver and Return processes, as some of them may not have been implemented by the node (e.g. the node is a distributor and therefore there is no Make process). Regarding the Return process, the configuration refers to whether the return flow is related to defective, Maintenance, Repair and Overhauling (MRO) or excess returns.

Level 3 of the SCOR model describes the individual steps required to complete each level-2 process. For each step, input and output data, along with a set of *best practices*[2] to carry out these steps and the required supporting capabilities are also defined in the model. An example can be found in Table 2.1, where the steps for the level-2 sub-process 'Make-To-Stock' are presented.

[2] Best practices are defined by the SCOR model as actual, structured practices that have proved to yield positive results for a host of industries, even if these results are not uniform across industries or SCs.

A fourth level is envisioned in the SCOR model. This level-4 model refers to the specific implementation of the SC activities and therefore is considered to be outside of the SCOR model, as it is not industry-neutral. Finally, performance indicators are attached in the SCOR model to each one of the activities/processes for each level (1–3), which provides a hierarchy of performance indicators that are (at least in principle) consistent among them.

Among the activities described in the SCORE model, there are two that have a great impact on the SC dynamic performance. These are demand modelling and forecasting (which is discussed in Sect. 2.3) and the management of the inventories (which is discussed in Sect. 2.4).

2.3 Demand Modelling and Forecast

Forecasting the customer demand is a fundamental task for every company involved in a SC. The basic goal of demand forecasting is to estimate the quantity of a product or service that the customers of a company (either the final customers or just another node downstream the SC) will require in the future. The importance of this business function is exacerbated when the node operates in an MTS environment, since the customer expects that the product or service is readily available, but it is also of utmost importance in an MTO environment, since ideally the raw materials and components of the product must be in the company's warehouses once the customer order arrives in order not to induce longer lead times. Even in an ETO environment, some demand forecast is required to make sure that the company has sufficient resources (personnel, machines, etc.) to face the customer demand.

Demand modelling differs from demand forecasting in the fact that the goal of the latter is to produce a model—hence the name—to express the evolution of the demand of a product across time. In the demand models that we address in the book (time series), different aspects of the demand such as the trend and seasonality are captured along with some aspects that cannot be captured (the so-called white noise), so a formula to express the stream of demand is obtained. Of course, these models can be used to forecast the demand, and indeed, instead of producing a single value of the future demand as done, e.g. in classical forecasting techniques, we can use the demand model to produce a range of possible outcomes, each one of them associated with a given probability.[3] However, in this book we will often use demand models to measure the performance of the SC depending also on the different aspects that make the stream of demand.

In the next sections, we will discuss several aspects related to demand modelling and forecasting that will be used profusely in the book. In Sect. 2.3.1, we present the characteristics of one specific time-series model that is widely in SCM, i.e. the so-

[3] Another reason frequently mentioned to use demand modelling techniques instead of forecasting techniques is their apparent better performance when dealing with non-aggregated product demand.

called AR(1) model. This model is suitable to express the demand for many products and therefore, it has been widely used to understand phenomena related to lead times, forecasting, inventory and the bullwhip effect.

2.3.1 Demand Modelling

Since the retailer does not know for sure what the future demand is, it is possible to think that D_t the demand on each period is a random variable. In this manner, we will have a time series of random variables. Throughout this book, we will assume that the demand follows the so-called first-order autoregressive model—or AR(1) model for short—where D_t the demand observed in a period can be expressed as (see Appendix C for further details on the AR(1) model):

$$D_t = d + \rho \cdot D_{t-1} + \epsilon_t \tag{2.1}$$

where $d \geq 0$ is a constant, ρ is the correlation constant ($|\rho| < 1$) and ϵ_t or *white noise* follows a normal distribution with mean zero and variance σ^2.

The correlation constant ρ expresses how the demand in the past periods is related to the demand in the next period. If $\rho = 0$, then there is no correlation between the demand among the periods (in other words, the demand is iid), while if $\rho > 0$, then the demand occurred in one period positively influences the next demand. In contrast, for $\rho < 0$, the demand between two consecutive periods is negatively correlated. Figure 2.3 shows some samples.

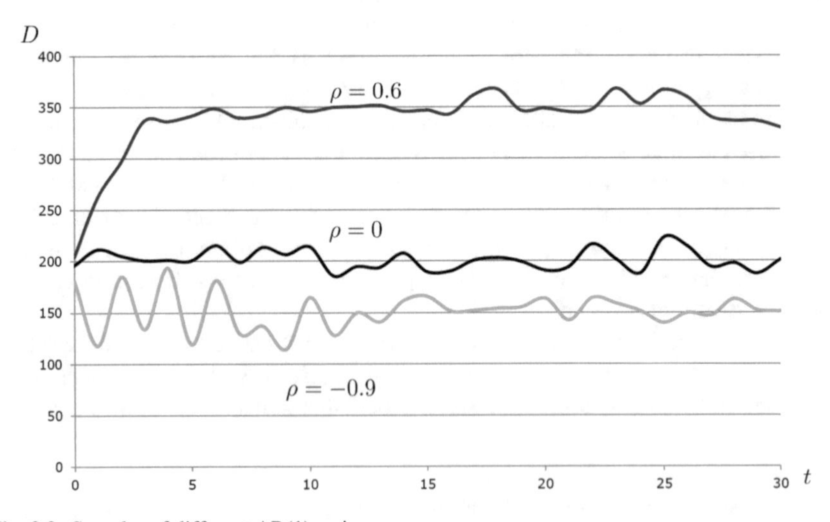

Fig. 2.3 Samples of different AR(1) series

Many final products fit into the AR(1). Furthermore, there is empirical evidence that many products from different sectors (ranging from the high-tech industry to groceries in the supermarkets) have $\rho > 0$. Therefore, their justification to model the customer demand is well-established.

AR(1) series have a number of properties that we will use in the models (their proofs are given in Appendix C):

- The mean of a demand expressed as an AR(1) series is $E[D] = \frac{d}{1-\rho}$.
- The variance of a demand expressed as an AR(1) series is $V[D] = \frac{\sigma^2}{1-\rho^2}$.
- The covariance between the demands occurring in two time periods separated by h period is $cov(D_t, D_{t+h}) = \rho \cdot V[D]$, regardless of the specific time t.

Finally, recall from Appendix C that it is sometimes convenient to write the AR(1) in an equivalent expression that is called the *causal form* of AR(1) (see Eq. (C.2) in Appendix C):

$$D_t = \frac{d}{1-\rho} + \sum_{i=0}^{\infty} \rho^i \cdot \epsilon_{t-i} = \mu + \sum_{i=0}^{\infty} \rho^i \cdot \epsilon_{t-i} \tag{2.2}$$

In most of the models of the book, we will assume that the demand follows the AR(1) model although, in some cases, in order to restrict the complexity of the resulting models, we will further assume that the demand is iidd, which—as stated before—is a particular case of the AR(1) model if $\rho = 0$.

2.3.2 Forecasting Techniques

In this section, we will present two widely established techniques for demand forecasting. The goal of a technique to forecast demand is to provide a close formula that we will use in time period t to estimate the future demand in period $t + h$ $(h > 0)$. We will denote such estimate as $\hat{d}_t(h)$. We will assume that, at time t, the actual demand data in $t, t-1, \ldots$ are known, i.e. we have $d_t, d_{t-1}, \ldots$.[4] In accordance with the conventions in the book, the d_t (known) demand at period t is denoted using lowercases whereas in the previous section D_t the demand in period t within a time series is denoted using uppercases as it is a RV.

More specifically, a forecasting technique will provide a function $f()$ such as

$$\hat{d}_t(h) = f(d_t, d_{t-1}, \ldots, h) \tag{2.3}$$

The simplest and most widely used forecasting techniques are the moving average and the exponential smoothing. Both are discussed in the next subsections.

[4] Note that this is not always the case in other texts dealing with this topic, where it is assumed that the last known demand data at time t is $t-1$. Of course, this does not affect the validity of the formulae shown below, but it modifies them as the summations involving the demand usually start one time period later.

2.3.2.1 Moving Average

The moving average (or simple moving average) forecasting technique predicts that the demand in any future period is given by the last m available demand data. More specifically, $\hat{d}_t(h)$ the forecast produced in time t regarding the future demand in period $t+h$ is

$$\hat{d}_t(h) = \frac{\sum_{i=0}^{m-1} d_{t-i}}{m} \tag{2.4}$$

As we can see, $\hat{d}_t(h)$ does not depend on h, which means that the forecasted demand (at time t) for all the future periods is the same; therefore, $\hat{d}_t(h)$ is usually denoted simply as $\hat{d}_t$ in the context of this forecasting technique. It can be seen then that this technique (or function) has only one *parameter*, i.e. m the number of time periods considered to compute the moving average. m serves to make the forecast more reactive or more conservative to changes in the demand: it is clear that the higher m, the less influence of a recent spike or fall in the forecast. In contrast, high values of m make the model less reactive to changes in the demand.

A potential problem of this method is that it gives equal weights to all used demand data. A generalisation of this technique is the so-called *weighted moving average*, which gives a different weight w_t to each data used in the formula:

$$\hat{d}_t(h) = \frac{\sum_{i=0}^{m-1} w_{t-i} \cdot d_{t-i}}{m} \tag{2.5}$$

This generalisation can be used, e.g. to give more weight to more recent data or to decrease the overall influence in the formula of a given period with abnormal data (due to, e.g. the existence of promotions, proximity of holidays, etc.)

2.3.2.2 Exponential Smoothing

Exponential smoothing is a forecasting technique that uses a weighted moving average of past data. The weights given to the past data are increasing with their recency, i.e. the highest weights are assigned to the most recent data. The relative value of these weights is controlled by the so-called *smoothing constant* $\alpha (0 < \alpha < 1)$, so $\hat{d}_t(h)$ the demand forecast produced in period t regarding the demand in $t+h$ is computed as follows:

$$\hat{d}_t(h) = \alpha \cdot d_t + (1-\alpha) \cdot \hat{d}_{t-1}(h) \tag{2.6}$$

As it can be seen, there is nothing in this technique that makes $\hat{d}_t(h)$ to be depending on h; therefore, we can write simply:

$$\hat{d}_t = \alpha \cdot d_t + (1-\alpha) \cdot \hat{d}_{t-1} \tag{2.7}$$

An alternative manner to write it is the following:

$$\hat{d}_t(h) = \hat{d}_{t-1} + \alpha \cdot \left(d_t - \hat{d}_{t-1}\right) \tag{2.8}$$

which expresses that the forecast produced in time t simply corrects the forecast produced in the previous period by a term that weights by α the error committed in the previous forecast (note that the error of the previous forecast is the difference between the actual demand in t and its forecast).

From Eq. (2.7), we can express $\hat{d}_{t-1}$ in a recursive manner, i.e. as a function of $\hat{d}_{t-2}$:

$$\hat{d}_t = \alpha \cdot d_t + (1-\alpha) \cdot \left(\alpha \cdot d_{t-1} + (1-\alpha) \cdot \hat{d}_{t-2}\right) = \tag{2.9}$$

$$\alpha \cdot d_t + \alpha \cdot d_{t-1} \cdot (1-\alpha) \cdot d_{t-1} + (1-\alpha)^2 \cdot \hat{d}_{t-2} \tag{2.10}$$

If we continue with the recursive expression, we would obtain

$$\hat{d}_t = \sum_{k=0}^{\infty} \alpha \cdot (1-\alpha)^k \cdot d_{t-k} \tag{2.11}$$

2.3.3 *Other Forecasting Techniques*

The techniques discussed in the previous sections are just the tip of the iceberg regarding forecasting techniques. Note that both the moving average and the exponential smoothing seem adequate whenever the demand is *levelled*, i.e. there is no increasing or decreasing trend of the demand. If this is not the case, it would be more convenient to use other techniques that explicitly consider this trend in the demand. Similarly, if the demand for the product is seasonal, there are techniques that can take this aspect into account.

Moreover, note that the two techniques do not seem to be very adequate to, e.g. forecast the demand of a certain spare part of a system which only happens to be seldom demanded. If we apply the moving average (or the exponential smoothing), we will find out that the demand is zero for most of the periods, and therefore, the forecast would be zero. In this case of *intermittent* demand, there are specific techniques that can yield very good results.

Finally, note that these techniques assume that we have consistent, reliable and historical data. This may not be the case if we try to forecast the demand for a new product. In this case, a very different technique must have to be used to provide a reliable forecast.

2.3.4 Measuring the Error in the Estimations

No matter what forecasting technique is used to estimate the demand, there is something that we can take for granted: there would be an error in the estimation, i.e. d_t the actual demand that occurred in time t would be different than its corresponding estimate $\hat{d}_t$. Clearly, we can only measure the error *after* the demand has occurred and, as a consequence, this error would be different in the future. However, it is of interest to measure this error to assert, on a statistical basis, what is the average error.

There are different indicators to measure the forecast error and, as we will see, none of them is entirely free of problems. Here, we present some of them:

- *Mean Error* or ME across m periods, which can be defined as follows:

$$ME = \frac{1}{m}\sum_{t=1}^{m}(d_t - \hat{d}_t) \tag{2.12}$$

  This measure of the error compensates for the under- and over-estimations in the forecast technique across the periods. Therefore, it is a good indicator to check whether the forecast method is biased, as a non-biased method should give an ME of zero or close to zero. Another problem of the method is that since the number of units of the demand on each period is not taken into account, then relatively accurate demand estimates of high demand may yield a high error whereas relatively inaccurate estimates of low demand can give a low error. This problem is avoided with the following indicator:

- *Mean Percentage Error* or MPE across m periods, which can be defined as follows:

$$MPE = \frac{1}{m}\sum_{t=1}^{m}\frac{d_t - \hat{d}_t}{d_t} \tag{2.13}$$

  MPE scales the size of the error in the forecast method with the numbers of units to be estimated. It can also be used to check the bias of the forecast method since it should be zero or close to zero if the method is not biased. Sometimes, the method is employed as a percentage, multiplying each term by 100. Note however that this indicator cannot be used if the demand in some period is zero, as it may happen in the case of lumpy or intermittent demand.

- *Mean Absolute Percentage Error* or $MAPE$ across m periods, which can be defined as follows:

$$MAPE = \frac{1}{m}\sum_{t=1}^{m}\frac{|d_t - \hat{d}_t|}{d_t} \tag{2.14}$$

  According to different studies, $MAPE$ is the most employed indicator to measure the accuracy of the demand forecast. In some cases, it is also expressed as a percentage, multiplied by 100.

As a rule of thumb, it is customary to classify the accuracy of a forecast according to the $MAPE$ values in the following categories:

- Accurate forecast: $MAPE < 10\%$.
- Good forecast: $MAPE$ between 10 and 20%.
- Reasonable forecast: $MAPE$ between 20 and 50%.
- Inaccurate forecast: $MAPE < 50\%$.

Despite its widespread usage, a problem of $MAPE$—which is also present in MPE—is the fact that it cannot be used if the demand can be zero. Another problem is that this method systematically gives better values if the method underestimates the demand than if it overestimates the demand.

- *Mean Absolute Error* or MAE across m periods, which can be defined as follows:

$$MAE = \frac{1}{m} \sum_{t=1}^{m} |d_t - \hat{d}_t| \tag{2.15}$$

As it can be seen, MAE is similar to $MAPE$ but the error is not normalised, so it can be used if the demand is zero in some period. In contrast, it has the same already-discussed problem as with ME regarding the size of the error.

- *Mean Squared Error* or MSE across m periods, which can be defined as follows:

$$MSE = \frac{\sum_{i=1}^{m} (d_t - \hat{d}_t)^2}{m} \tag{2.16}$$

As it can be seen, MSE measures the average discrepancy between the estimation and the actual demand (squared so that the over- and underestimates are not cancelled). This is another frequently used measure of the forecast error, which is similar to $MAPE$ or MAE, but it has the advantage that it uses a 'nicer' function than the absolute value. A problem associated with MSE is that the error does not have the same units as the demand.

- *Root of the Squared Error* or $RMSE$ across m periods, which can be defined as follows:

$$RMSE = \sqrt{\frac{\sum_{i=1}^{m} (d_t - \hat{d}_t)^2}{m}} \tag{2.17}$$

In this indicator, the error has the same units as the demand.

2.3.5 MMSE Estimation

From the previous comments, it is clear that any estimation of the demand would imply an error in an estimation, and that this error can only be measured a posteriori in the sense that a forecasting method showing a low error in the past does not guarantee the same performance in the future. In this situation, one question that is relevant is the following: What is the best that we can expect from *any* forecasting technique? In other words, what would be the minimum error that one can expect in the long run? From time-series theory (see Appendix C), we know that, for a given time-series model of the demand, we can derive an estimator that minimises the MSE indicator. Note that such MMSE estimation is not a forecasting technique *per se* as the moving average or the exponential smoothing, but a tool that allows us to guess which prediction we would make under optimal conditions (again, in the sense that MSE is minimised on the long run).

More specifically, from time-series theory, if we have the following time series of the demand (written in its causal form):

$$D_t = \mu + \sum_{i=0}^{\infty} \psi_i \cdot \epsilon_{t-i} \tag{2.18}$$

then the 'best' estimator for the demand at time $t + h$ that we can build at time t in the sense that this estimator would minimise the MSE in the long run (MMSE estimator) is the following:

$$\hat{D}_t(h) = \mu + \sum_{i=0}^{\infty} \psi_{h+i} \cdot \epsilon_{t-i} \tag{2.19}$$

and that the value of $e_t(h)$ the error committed in the forecast made in period t regarding the demand in $t + h$ is (see Eq. (C.36)) as follows:

$$e_t(h) = D_{t+h} - \hat{D}_t(h) = \sum_{i=0}^{h-1} \psi_i \epsilon_{t+h-i} \tag{2.20}$$

For the case of an AR(1) time series, from the comparison of Eq. (2.18) with Eq. (2.2), we can see that $\psi_i = \rho^i$; therefore, the MMSE estimator is

$$\hat{D}_t(h) = \mu + \sum_{i=0}^{\infty} \rho^{h+i} \cdot \epsilon_{t-i} \tag{2.21}$$

Note that Eq. (2.21) can be written as

$$\hat{D}_t(h) = \mu + \rho^h \sum_{i=0}^{\infty} \rho^i \cdot \epsilon_{t-i} \tag{2.22}$$

and that the sum in the last term of Eq. (2.22) simply represents the causal form of the AR(1) model, see Eq. (C.2). Therefore, we can write

$$\hat{D}_t(h) = \mu + \rho^h (D_t - \mu) \tag{2.23}$$

or, if we use the equality $\mu = \frac{d}{1-\rho}$,

$$\hat{D}_t(h) = \frac{d}{1-\rho} + \rho^h \left(D_t - \frac{d}{1-\rho} \right) \tag{2.24}$$

On the other hand, the value of the error is

$$e_t(h) = \sum_{i=0}^{h-1} \rho^i \epsilon_{t+h-i} \tag{2.25}$$

We will use these results in several chapters, particularly when we assume that the demand is iid (or, in other words, that $\rho = 0$ in the AR(1) model). In this case, the MMSE estimator is simply $\hat{D}_t(h) = \mu$, that is, when the demands are iid, then the MMSE estimation of the future demand is the average demand.

It is perhaps interesting to investigate the expected value and the variance of the error of the MMSE estimation for the AR(1) models. Regarding the expected value, it is straightforward to see that $E[e_t(h)] = 0$ since by hypothesis $E[\epsilon_t] = 0$. With respect to the variance, we obtain the following expression:

$$V[e_t(h)] = \sum_{i=0}^{h-1} \left(\rho^i\right)^2 V[\epsilon_{t+h-i}] = \sum_{i=0}^{h-1} \rho^{2i} \cdot \sigma^2 = \frac{1-\rho^{2h}}{1-\rho^2} \cdot \sigma^2 \tag{2.26}$$

For the case of iid demand, we see that the variance of the error is $V[e_t(h)] = \sigma^2$, i.e. it is the same as that of the white noise. For the general case, we see that the variance of the error increases with h, i.e. the further the future demand we wish to predict, the higher the variability of the error.

2.3.5.1 Estimating the Total Demand Across Several Periods

As we will see in Sect. 2.4, sometimes it is required to forecast the total demand that will occur in the next L time periods ($L \geq 1$). In this book, such need will be motivated by the fact that the orders issued to the provider at time t will, in general, arrive $t + L$ periods later and therefore, the node needs to estimate the demand that he/she will face during these L periods. This period is known as *risk period* in inventory management theory. In this section, we will give the expressions of the estimates of the total future demand depending on the forecasting method.

More specifically, in time period t, we wish to obtain $\hat{D}_t^L$ the estimation of the total demand occurring during the next L periods, i.e.

$$\hat{D}_t^L = \hat{D}_t(1) + \hat{D}_t(2) + \cdots + \hat{D}_t(L) = \sum_{h=1}^{L} \hat{D}_t(h) \tag{2.27}$$

Note that the first estimation is performed in time $t+1$, as we have assumed throughout this book that the actual demand in period t is known at the time that any forecast is done.

Depending on the forecast technique employed, Eq. (2.27) takes different expressions. If a moving average forecasting technique with m periods is employed, then we have

$$\hat{D}_t^L = \sum_{h=1}^{L} \hat{D}_t(h) = \sum_{h=1}^{L} \left(\frac{\sum_{i=0}^{m-1} D_{t-i}}{m} \right) = \frac{L}{m} \cdot \sum_{i=0}^{m-1} D_{t-i} \tag{2.28}$$

If an exponential smoothing technique is employed, then the expression is

$$\hat{D}_t^L = \sum_{h=1}^{L} \hat{D}_t(h) = \sum_{h=1}^{L} \sum_{i=0}^{\infty} \alpha(1-\alpha)^i D_{t-i} = L \cdot \sum_{i=0}^{\infty} \alpha(1-\alpha)^i D_{t-i} \tag{2.29}$$

Finally, if a MMSE estimation is employed and the demand is, as we have usually assumed, AR(1), then we can obtain a more involved expression by substituting the single-period MMSE estimation in Eq. (2.21):

$$\hat{D}_t^L = \sum_{h=1}^{L} \hat{D}_t(h) = \sum_{h=1}^{L} \left(\mu + \sum_{i=0}^{\infty} \rho^{h+i} \cdot \epsilon_{t-i} \right) = \tag{2.30}$$

$$L \cdot \mu + \sum_{h=1}^{L} \sum_{i=0}^{\infty} \rho^{h+i} \cdot \epsilon_{t-i} = L \cdot \mu + \sum_{h=1}^{L} \rho^h \left(\sum_{i=0}^{\infty} \rho^i \cdot \epsilon_{t-i} \right) = \tag{2.31}$$

$$L \cdot \mu + \frac{\rho - \rho^{L+1}}{1-\rho} \sum_{i=0}^{\infty} \rho^i \cdot \epsilon_{t-i} \tag{2.32}$$

It can be seen that the last summation can be written simply as $D_t - \mu$, since this is precisely the causal expression of the AR(1) demand. See Eq. (2.2). As a consequence, we have

$$\hat{D}_t^L = L \cdot \mu + \frac{\rho - \rho^{L+1}}{1-\rho} \cdot (D_t - \mu) \tag{2.33}$$

It is useful also to compute e_t^L the cumulative error if using the expression in (2.33) to estimate the total demand in the L periods, i.e. the sum of the errors committed across the L periods:

$$e_t^L = \sum_{h=1}^{L} e_t(h) = \sum_{h=1}^{L} \sum_{i=0}^{h-1} \rho^i \epsilon_{t+h-i}$$

or

$$e_t^L = \underbrace{\epsilon_{t+1}}_{h=1} + \underbrace{\epsilon_{t+2} + \rho\epsilon_{t+1}}_{h=2} + \underbrace{\epsilon_{t+3} + \rho\epsilon_{t+2} + \rho^2\epsilon_{t+1}}_{h=3} + \cdots + \underbrace{\epsilon_{t+L} + \rho\epsilon_{t+L-1} + \cdots + \rho^{L-1}\epsilon_{t+1}}_{h=L} =$$

$$= (1 + \rho + \cdots + \rho^{L-1})\epsilon_{t+1} + (1 + \rho + \cdots + \rho^{L-2})\epsilon_{t+2} + \cdots + \epsilon_{t+L} = \sum_{i=0}^{L-1} \epsilon_{t+L-i} \left(\sum_{j=0}^{i} \rho^j \right)$$

which results in the following expression for e_t^L:

$$e_t^L = \frac{1}{1-\rho} \sum_{i=0}^{L-1} (1 - \rho^{i+1})\epsilon_{t+L-i} \tag{2.34}$$

Since we know that $e_t \sim N(0, \sigma^2)$, it is straightforward to see that $E[e_t^L] = 0$. Regarding the variance and taking into account the iid nature of ϵ_t, we have

$$V[e_t^L] = \sum_{i=0}^{L-1} \sigma^2 \left(\frac{1 - \rho^{i+1}}{1 - \rho} \right)^2 = \frac{\sigma^2}{(1-\rho)^2} \left(L + \sum_{i=0}^{L-1} \rho^{2(i+1)} - 2 \sum_{i=0}^{L-1} \rho^{i+1} \right)$$

which results in the following expression for the cumulative error:

$$V[e_t^L] = \frac{\sigma^2}{(1-\rho)^2} \left(L + \rho^2 \frac{1 - \rho^{2L}}{1 - \rho^2} - 2\rho \frac{1 - \rho^L}{1 - \rho} \right) \tag{2.35}$$

Equation (2.35) would be used in the subsequent chapters, due to the following interesting property: It can be seen that $V[e_t^L]$ does not depend on t, which means that the variance of the cumulative error in the demand forecasting is constant over time. Since, as we will see, the standard deviation of such cumulative error is employed in some replenishment policies discussed in Sect. 2.4—most notably the OUT policy widely used in the remaining chapters of the book, where it is used to compute the required safety stock—we will use this result to justify that it seems reasonable to use the same safety stock for all periods, at least within the hypotheses of the expression (MMSE estimation of an AR(1) demand).

2.4 Source: Inventory Management

The management of the inventories in a company is central to its operations. Inventories are constituted by products at some stage of their processing (raw materials, semi-finished products and final products); therefore, the company has purchased them and/or has devoted some resources to manufacture them or simply to keep them stored. This makes inventories expensive and, idealistically, the best option for the nodes in a SC would be not to have them. However, this is not an option in most cases, since one or more of the following situations may occur:

- The demand for the final products is not known in advance, and the customer is not willing to wait until the raw materials are purchased and the product is processed on the shop floor.
- Companies have a finite production capacity, which in turn is usually expensive and as a consequence quite adjusted to the (average) expected demand; therefore, if the demand of a product is highly seasonal, then the logical solution is to produce more than it is demanded in off-peak times in order to be able to fulfil the demand during peak times.
- There are usually costs associated with issuing or fulfilling an order, ranging from administrative processing costs to transportation costs. These costs are fixed or, at best, not linear (it is clear, e.g. that the fuel cost of a half-empty track is roughly the same as those of a full-truck). This makes it un-economical to place an order every time a unit of a product is needed. Furthermore, volume discounts are quite common for many products, thus making it more profitable to place bigger orders less frequently.
- Some products may not simply be available in exactly the required quantity, as they are sold in pallets, cases, containers, etc., which results in the need of rounding up the required quantity and storing the leftovers.
- Providers are not always reliable and, like any company, can have capacity problems and delays to serve the required material. A natural way to protect the node against these fluctuations and unreliability of the demand is to maintain a stock of raw materials and components of the product.

The situation described above, among others, speaks for the need of maintaining inventories across the whole SC and in every stage of the processing of a product. As we can see, in addition to the costs associated with having the inventory, there are costs associated with not having enough inventory. As a consequence, most inventory management evolves around the idea of calculating the amount of inventory that minimised the total costs. It is perhaps useful then to briefly discuss these costs, as they will appear in the models developed later in the section. These are as follows:

- Fixed costs. This is the cost to place an order, regardless of its size. The fixed costs typically include the administrative and transportation costs to place an order. We will assume that these fixed costs are K.
- Inventory handling costs. This represents the cost of keeping the inventory at hand (cost of the storage space, personnel and equipment to handle the inventory, insurance, etc.). We will assume that it is possible to assign handling cost h to each unit of product in the inventory per unit time.
- Stockout costs. This is the cost of not having sufficient inventory to fulfil the customer demand. If the customers are willing to wait until their requests are met, then these costs may include the potential loss of future orders since the customers need have not been met and—in some cases—a penalty or compensation that it

may be imposed by the customer. If the customers are not willing to wait, then these costs should include the loss of profit due to the missed order. We will assume that it is possible to assign a cost b per product unit whose demand has not been met.

Therefore, the basic problem of inventory management that we will address is to develop some rules (or *policies*) to decide when to order and what quantity has to be ordered so the total costs are minimised. Clearly, these two decisions are related and are usually addressed by using **inventory models** that formulate this generic decision problem under different assumptions (the nature of the demand, the lead times, etc.) and find its optimal solution.

There is a broad distinction between the deterministic inventory models (where it is assumed that the demand is known in advance) and stochastic inventory models (where the demand is assumed to be a RV). In this book, we will focus on stochastic inventory models, as they produce the most common and most noxious effects on the SC performance.

In these models, we typically consider the following magnitudes that evolve around time:

- *On-hand* inventory. This is the physical inventory of the product and it is usually denoted by I. Whenever there is no confusion with other terms defined below, we will denote it simply as inventory.
- Backlog. The backlog is the customer demand that has not been met and is waiting to be fulfilled. It is denoted in the remainder of the section as B.
- Net inventory. In the case that the customer is willing to wait, the net inventory is defined as the physical inventory minus the backorder units. Therefore, the net inventory represents the physical inventory 'available' to meet new demands from customers.
- Inventory position. The inventory position is the inventory that is owned by the firm but that is not necessarily in its warehouses. More specifically, the inventory position is the net inventory plus the replenishment orders that have been placed but that have not arrived yet (these are also known as work in process) minus the backlog (demand that has occurred in the past but that has not been possible to meet). In other words, the inventory position is the *projected* inventory available to meet new customer demands. We can write

$$IP = I + O - B \tag{2.36}$$

where IP denotes the inventory position and O the orders placed but that have not arrived yet.

In stochastic inventory models, the demand is assumed to be a RV. To increase the mathematical tractability of these models, in most cases, the demand—which, in general, would consist of discrete units of a product—is modelled using a *continuous*, non-negative RV.

More specifically, we will be interested in a RV that measures the demand that occurs during L time periods. In the following, we denote by X such RV, which is characterised by a pdf $f(x)$, with $E[X] = \int_0^\infty x \cdot f(x)dx = \mu^L$ and $V[X] = \left(\sigma^L\right)^2$. As usual, the cdf of this RV is denoted by $F(x) = \int_0^\infty f(x)dx$.

Here, we will focus on the base stock model or Order-Up-To (OUT) policy. The main hypothesis that we consider are the following:

- There are no fixed costs, i.e. $K = 0$.
- Unfilled demand is backlogged, i.e. the customer is willing to wait until his/her demand is fulfilled. However, this comes at a cost (*backorder costs*) that we will have to quantify. As previously discussed, the cost of a backlogged unit is assumed to be b.
- Replenishment lead times are fixed.

In the base stock system, the inventory position is continuously monitored, and the idea is to keep the inventory position constant at a level s—denoted as base stock level. In other words, in a base stock system, we have that $s = I + O - B$. In this system, backorders will be generated if the inventory position at the time that an order is placed (this order will arrive after L time periods) is not enough to face the demand that will occur during these L time periods, i.e.

$$B(s) = \begin{cases} 0 & if X \leq s \\ X - s & if X > s \end{cases} \tag{2.37}$$

Alternatively, on-hand inventory will be generated if the inventory positions exceeds the demand that will occur during these L time periods, i.e.

$$I(s) = \begin{cases} s - X & if X \leq s \\ 0 & if X > s \end{cases} \tag{2.38}$$

The relationship between the expected values of $B(s)$ and $I(s)$ can be obtained as follows. The expression of $E[B(s)]$ is

$$E[B(s)] = \int_0^\infty B(s)dx = \int_s^\infty (x - s) \cdot f(x)dx = \tag{2.39}$$

$$\int_s^\infty x \cdot f(x)dx - s\int_s^\infty f(x)dx = \int_s^\infty x \cdot f(x)dx - s\left(1 - F(s)\right) \tag{2.40}$$

whereas the expression of $E[I(s)]$ is

$$E[I(s)] = \int_0^\infty I(s)dx = \int_0^s (s - x) \cdot f(x)dx = \tag{2.41}$$

$$s\int_0^s f(x)dx - \int_0^s x \cdot f(x)dx = sF(s) - \left(\mu - \int_s^\infty x \cdot f(x)dx\right) = \tag{2.42}$$

$$sF(s) - \mu + \int_s^\infty x \cdot f(x)dx = sF(s) - \mu + E[B(s)] + s - sF(s) = \quad (2.43)$$

$$s - \mu + E[B(s)] \quad (2.44)$$

As a consequence, $C(s)$ the function of the total (backlog plus inventory holding) expected costs depending on the base stock level s can be expressed as follows:

$$C(s) = h \cdot E[I(s)] + b \cdot E[B(s)] = h(s - \mu) + (h + b)E[B(s)] \quad (2.45)$$

The value of s that minimises the total expected costs (i.e. s^*) can be obtained by deriving the expression in Eq. (2.45) with respect to s and making this expression equal to zero. The derivative is

$$\frac{\partial C(s)}{\partial s} = h + (h + b)\frac{\partial E[B(s)]}{\partial s} \quad (2.46)$$

So the following equation must hold:

$$h + (h + b)\frac{\partial E[B(s)]}{\partial s} = 0 \quad (2.47)$$

We can obtain a more compact expression by computing $\frac{\partial E[B(s)]}{\partial s}$:

$$\frac{\partial E[B(s)]}{\partial s} = \frac{\partial}{\partial s}\left(\int_s^\infty x \cdot f(x)dx - s\,(1 - F(s))\right) = \quad (2.48)$$

$$\frac{\partial}{\partial s}\int_s^\infty x \cdot f(x)dx - (1 - F(s)) + s\frac{\partial F(s)}{\partial s} \quad (2.49)$$

To continue is useful to recall from Leibniz's rule that the first term is equal to $-s \cdot f(s)$ (see Appendix A.3) and that $\frac{\partial F(s)}{\partial s}$ is simply the pdf $f(s)$, therefore:

$$\frac{\partial E[B(s)]}{\partial s} = F(s) - 1 \quad (2.50)$$

Substituting in Eq. (2.47), we have that for s^* the optimal stock-base level, the following equation must hold:

$$h + (h + b)\big(F(s^*) - 1\big) = 0 \quad (2.51)$$

Therefore,

$$F(s^*) = \frac{b}{b + h} \quad (2.52)$$

In other words, s^* must be the integer that makes the cdf of the demand to have the value $\frac{b}{b+h}$ or

$$P[X \leq s^*] = \frac{b}{b+h} \tag{2.53}$$

If we can assume that the demand that occurs during the L periods is normally distributed (with mean μ^L and standard deviation σ^L), then the above expression can be standardised, i.e.

$$P\left[\frac{X-\mu^L}{\sigma^L} \leq \frac{s^*-\mu^L}{\sigma^L}\right] = \frac{b}{b+h} \tag{2.54}$$

and, if z is the value that obtain a probability $\frac{b}{b+h}$ in the standard normal distribution, then if follows that

$$s^* = \mu^L + z \cdot \sigma^L \tag{2.55}$$

The term $z \cdot \sigma^L$ in Eq. (2.55) is sometimes referred to as safety stock since in this policy we intend to face the demand during the next L periods using the inventory position s. Therefore, if the demand is what it is expected, then $s - \mu^L = 0$. The difference $s - \mu^L$ is used to protect against stockouts due to the fluctuation in the demand. Similarly, z is denoted as safety factor and here its optimal value has been obtained taking into account the minimisation of the total costs.

Note that an alternative manner to obtain z would be to specify a desired customer service level or fill rate f. The fill rate is simply defined as the ratio of the demand that can be filled from stock, and it seems to be a reasonable indicator for customer service. Note that, under the OUT policy, this ratio is given by the probability that the demand does not exceed the base stock level; therefore, the minimal base stock level s that achieves the desired fill rate f is given by

$$P[X \leq s] = f \tag{2.56}$$

Again, if we assume that X is a normal RV, s can be obtained as

$$s = \mu^L + z_f \cdot \sigma^L \tag{2.57}$$

where z_f is the value in the standard normal distribution yielding a probability of f.

2.4.1 Forecasting the Demand

As we have seen in the previous section, if the demand is iid normally distributed and we use an OUT policy, we can compute the optimal stock level s^* using Eq. (2.55). To do so, in addition to setting the safety factor either according to the cost structure or to the desired service level, we need μ^L and σ^L the mean and standard deviation of the demand across the lead time L. Unfortunately, in most of the real-life settings, we don't know their *true* values so a forecast of both magnitudes has to be performed, so the base stock level would be determined using these estimates. With respect to an estimate of the mean demand, obviously, we can use any of the forecasting techniques discussed in Sect. 2.3.2 such as, e.g. the moving average. Regarding an estimate of the standard deviation, we could, in principle, use the sample variance of the demand as an estimator, i.e. use the following formula as estimate:

$$\hat{\sigma} = \sqrt{\frac{1}{m-1}\sum_{i=0}^{m-1}(d_{t-i} - \hat{\mu}_{t-i})^2} \tag{2.58}$$

However, it turns out that this is not the right standard deviation to use in Eq. (2.57). The reason is a bit complex so here we will provide only an intuitive explanation: As we discussed earlier in the context of the base stock policies, if the demand is what it is expected (or, to be precise, the actual demand coincides with the forecasted demand), then it should be ideally covered with s. The safety stock should then protect against the discrepancies between the actual demand and the forecasted demand, i.e. the forecast error. More specifically, it is the variance of the forecast error that causes the problem (and the phenomenon that we should protect against), since a consistent bias in the forecast error can be corrected simply by adjusting the forecast, whereas a variance in this error shows the inability of the forecast method to accurately estimate the demand.

As a consequence, it is the standard deviation of the forecast error that should be integrated into Eq. (2.57) when the average demand is not known (i.e. in practically all real-life settings). This distinction is somehow important, but it should not be overestimated: in many settings, there is little evidence that the demand is iid normally distributed and, if this is not the case, Eq. (2.57) is no longer an optimal expression, but an approximation. Furthermore, the proportionality of the standard deviation of the error and that of the demand can be established for certain forecasting techniques (including the moving average and the exponential smoothing); therefore, the error induced can be subsumed in z.

2.4.2 *Estimating the On-Hand Inventory*

In this section, we will develop an approximation of the average on-hand inventory ($\bar{I}$) when using the OUT policy that it will be used later in the book (more specifically, in Chap. 5). Here, we will provide an intuitive explanation taking into account the following aspects:

- We will first assume that D the demand in each period is deterministic. As we will see, this hypothesis simplifies obtaining an expression for $\bar{I}$ but at the same time converts the so-obtained expression into an approximation. We will discuss later the accuracy of the approximation.

- In order to keep the inventory position constant (equal to s), the order O issued at a period should compensate the demand that has occurred during this period, i.e. $O = D$.
- Since the inventory is continuously monitored and the demand is deterministic, the lowest level of the on-hand inventory will occur just before the arrival of the order, where the remaining inventory is just the safety stock ($z\sigma^L = s - \mu^L$), and the highest level will occur just after the arrival of an order O (in this moment, the on-hand inventory would be $z\sigma^L + D$). As a consequence, the average inventory would be

$$\bar{I} = \frac{(z\sigma^L + D) + z\sigma^L}{2} = z\sigma^L + \frac{D}{2} = \frac{D}{2} + s - \mu^L \tag{2.59}$$

In reality, the demand is not deterministic and this can cause backorders to occur. Furthermore, the demand for each period would different, so we may substitute D by the expected value of the demand $E[D_t]$, and μ^L the expected demand across the lead time by $E[D_t^L]$. Finally, the safety stock may be different for each period, so Eq. (2.59) can be rewritten as the following approximation:

$$\bar{I} \approx \frac{E[D_t]}{2} + s_t - E[D_t^L] \tag{2.60}$$

This approximation is widely used and considered to be excellent when the backorder costs are substantially higher than the inventory holding costs. In order not to be considered an approximation, the formula should take into account the fact that backorders can occur, so it is exact if a last term with the average backorder level is included.

2.5 Summary

In this chapter, we have focussed on the main decisions involving the operation of a typical SC. These encompass the forecasting or modelling of the demand in order to anticipate the final customer, and the inventory management and replenishment decisions. Regarding demand forecasting, we have focussed on two relatively simple (but widely used) techniques, i.e. the moving average and the exponential smoothing. Regarding demand modelling, we have discussed the AR(1) time series for modelling the final customer demand. Regardless of the technique used to forecast the demand, a forecast error is always introduced, and it is relevant to measure it (using different indicators) in order to quantify the (past) errors committed by a forecasting technique. Equally interesting is to measure the error committed if using the MMSE estimation, which can be seen as a proxy for the best technique (whatever it is) that we could hope to forecast the demand. We will use extensively the MMSE estimation in the next chapters as a benchmark. The last part of the chapter involves some basics of inventory management, particularly regarding the base stock or OUT policy, which is optimal under certain conditions (if the demand is iid normally distributed), and again, it would be used as a benchmark replenishment policy. After this chapter, we are now equipped with the basic ingredients to model the different elements of a supply chain. In the next chapter, we will discuss the dynamic phenomena that we intend to model and which are the most suitable indicators to measure them.

2.6 Further Readings

The description of standard forecast techniques such as the moving average or the exponential smoothing can be found in many textbooks, such as in [1] or in [2]. A relatively recent contribution on the role of forecasting techniques in SCs is [3]. Evidence regarding the suitability of AR(1) time series to model the demand can be found in different papers. Also, the statement that most products have $\rho > 0$ can be found in [4] after studying the sales pattern of 150 different SKUs in a supermarket. Reference [5] found that it is common to have positive correlations with values as high as $\rho = 0.7$ in high-tech and product consumer industries. Reference [6] found also positive correlated values for supermarket products.

A detailed proof on the suitability of using the standard deviation of the forecasted error instead of the standard deviation of the demand is provided in [7], and the intuitive explanation is given in [1]. The equivalence (under certain conditions) of these two magnitudes is given in [8]. The approximation of the average on-hand inventory discussed in Sect. 2.4.2 can be found in [9] or in [10], and its suitability is discussed in [4]. There are other, more precise approximations for the average on-hand inventory in [11], although these methods do not result in a close form expression.

References

1. Snyder, L., Shen, Z.M.: Fundamentals of Supply Chain Theory, 2nd edn. (2019)
2. Shapiro, J.F.: Modeling the Supply Chain. Duxbury (2001)
3. Syntetos, A., Babai, Z., Boylan, J., Kolassa, S., Nikolopoulos, K.: Supply chain forecasting: theory, practice, their gap and the future. Eur. J. Oper. Res. **252**(1), 1–26 (2016)
4. Lee, H., So, K., Tang, C.: Value of information sharing in a two-level supply chain. Manag. Sci. **46**(5), 626–643 (2000)
5. Erkip, N., Hausman, W.H., Nahmias, S.: Optimal centralized ordering policies in multi-echelon inventory systems with correlated demands. Manag. Sci. **36**(3), 381–392 (1990)
6. Ali, M., Boylan, J., Syntetos, A.: Forecast errors and inventory performance under forecast information sharing. Int. J. Forecast. **28**(4), 830–841 (2012)
7. Nahmias, S.: Production and Operations Analysis. McGraw-Hill, New York (2005)
8. Hax, A., Candea, D.: Production and Inventory Management. Prentice-Hill, Hoboken (1984)
9. Silver, E., Peterson, R.: Decision Systems for Inventory Management and Production Planning (1985)
10. Hopp, W., Spearman, M.: Factory Physics (1996)
11. Zipkin, P.: Foundations of Inventory Management (2000)

Chapter 3
Supply Chain Dynamics

3.1 Introduction

As we have discussed in Chap. 1, supply chain has become a prevalent form of producing and delivering items due to several advantages. However, there are numerous empirical evidence that many companies that operate in a supply chain experience periods where they have inventory shortages, followed by periods with inflated inventories. Clearly, extra costs are incurred in both periods, since shortages affect the service level committed by the company, while excess inventory represents additional costs in terms of inventory handling (and, in some cases, additional costs if the product is perishable). The costs associated with these fluctuations may vary from one industry to another, but some studies estimate excess costs around 12–25%.

Of course, there may be different reasons explaining such problems. In some companies, the lack of knowledge of the demand pattern might lead to poor forecasts, or the replenishment policies might not be adequate to cope with the demand instability. Unreliable providers or third-party logistics might also be blamed in some companies for a poor service level. However, the almost-ubiquity of this phenomenon led researchers to think that it can have some structural causes rooted in the very nature of the SC, as the recollection of this effect was recognised in diverse markets, ranging from diapers to printers or from the pharmaceutical industry to the computer market.

Among the first researchers to acknowledge this fact was Jay Forrester, who pointed out in the 60s that this effect (which he called 'demand amplification') was a consequence of industrial dynamics or time-varying behaviours of industrial organisations. In other words, the so-called bullwhip effect was a side effect of the way to organise the SC and not simply the result of the company- or market-specific ill choices.

From Forrester's pioneering work, many researchers have studied the bullwhip effect and identified its causes, have developed indicators to detect and measure it and have provided some remedies to mitigate it. In this chapter, we summarise these efforts in order to provide a context for the models that we will develop in the next chapters. More specifically, we:

J. M. Framinan, *Modelling Supply Chain Dynamics*,
https://doi.org/10.1007/978-3-030-79189-6_3

- Describe the dynamic behaviour of the SC, identifying different causes for the bullwhip effect (Sect. 3.4).
- Present the main indicators used to measure the bullwhip effect (Sect. 3.6).
- Discuss the approaches that have been tried to mitigate the bullwhip effect (Sect. 3.5).

3.2 The Dynamic Behaviour of a SC

The concept of the bullwhip effect first appeared in Jay Forrester's Industrial Dynamics in the 60s. It is mentioned in the context of a production-distribution system where delays and ordering policies, among others, play a key role in exacerbating demand variability. Building upon the industrial dynamics work of Forrester, a series of experiments in the 80s—using the Beer Distribution Game that is described below—illustrated that the bullwhip effect may be rooted in SC managers' misperceptions regarding demand, inventories and work in process (or pipeline inventory), thus making a case for *behavioural* causes of the bullwhip effect. In other words, the bullwhip effect was a consequence of the systematic irrational (i.e. non-rational) behaviour of the managers. These conclusions have not been challenged today, but instead, different behavioural causes of the bullwhip effect have been added to the list by different researchers, thus establishing a stream of causes labelled as behavioural.

However, managers' behaviour is not the only cause of the bullwhip effect. In a series of papers in the 90s, Hau Lee and other researchers developed several simple models to illustrate situations in which the bullwhip effect may exist if the decisions to be taken are not only purely rational, but optimal or pseudo-optimal. Interestingly, in some instances (such as in the case of price fluctuations), it was precisely the rational (optimal) behaviour of all parties involved the reason for the bullwhip effect to manifest. These causes are labelled as operational since they are rooted in the (typical) SC infrastructure.

Both causes (behavioural and operational) are not mutually exclusive and can be present in the same SC. However, it is interesting to separate them to classify and analyse them in detail. This is done in the next sections.

3.3 Behavioural Causes

It has long been understood that decision makers do not operate in an optimal (or rational) manner, and there are numerous field experiments proving that there are psychological biases affecting the decisions as well as cognitive limits regarding the data that humans may apprehend and process. Regarding those affecting the bullwhip effect, the following have been identified:

- **Overstatement of anecdotal data**. Some managers tend to overstate some anecdotal data. For instance, a customer complaining about his/her perceived service level may induce the manager to negatively assess the global service level of the company, even if the complaint does not come from a major customer and the overall service level may be regarded as acceptable. As a consequence, the manager might build up more inventory based on this individual perception.
- **Under- or over-reaction to demand changes**. As the demand is, in general, not fully stable, changes in the demand during consecutive periods can be seen as a *noise* within an overall levelled demand, or they can be read as the beginning of a change in the demand tendency. These different interpretations may lead to under- and over-estimations of the demand, which in turn would influence replenishment decisions. More generally, demand forecasting (which has been discussed in detail in Chap. 2) is inherently associated with a forecasting error, and when the realised demand does not match the forecast, there is the temptation to *tune* the next forecast by incorporating the past error. Unfortunately, this practice is likely to amplify the forecast error in the future.
- **Risk aversion**. There are empirical studies suggesting that risk-averse managers display hoarding behaviour when taking replenishment decisions in order to reduce the risk of running out of stock. This makes sense since, in most cases, the shortage costs are substantially higher and inventory holding costs.
- **Cognitive limitations**. Some supply chains may be extremely complex, operating with different products across a number of channels for different customer segments. Therefore, even for the most experienced managers, it might not be easy to grasp the full implications of their decisions. The structure of the SC may be extremely complicated to apprehend in full detail. A side effect is the fact that SC managers do not use all the available information in making their decisions. One of the often overlooked issues is the role of the orders in the pipeline (orders issued in the past but that have not arrived yet). An example is when, upon an expected future demand peak, the retailer issues orders to her supplier in order to adequately replenish the inventory. Since the time from issuing the orders to the supplier until the actual arrival of the product is not negligible, the orders are issued sometime in advance. This time in advance may be the result again of some forecast performed by the retailer in view of historical data, but in reality, it may be highly variable depending on the workload of the supplier (which it turn it may also depend on the workload of the supplier's suppliers), something that the retailer is often quite unaware. Therefore, if the arrival of the product is delayed, the peak of the demand has not been met (and backlog or lost sales costs have been incurred) at, when a few periods later, the products finally arrive, there is an excess inventory which cannot be used, thus incurring also in unnecessary inventory holding costs.

Surely the best-known illustration of the behavioural causes of the bullwhip effect is the *beer distribution game*. This is a simulation game that was initially developed by Jay Forrester in the 60s and then further developed by other researchers at MIT. Although there are many flavours and extensions of this game, in its original version, it has four players (nodes) representing four nodes of a beer supply chain (hence the name): the retailer, the wholesaler, the distributor and the manufacturer. One year (52 weeks) is simulated in this game and each week the retailer (or wholesaler, distributor) places an order to the wholesaler (distributor or manufacturer) and receives the result of the previous order from the corresponding player. Provided that a player has the inventory to satisfy the demand of his/her customer, there is a two-week delay from the reception of the order until the product physically arrives at the customer. The goal of the orders is to replenish the inventory to satisfy the demand from each node's customer, being the demand for the retailer generated exogenously (each week the retailer draws the current demand from a deck of cards). There are costs associated both with the inventory of each node and with backlog costs, so the goal of the players is to operate this supply chain with the lowest costs. There is no communication between the players, apart from receiving the orders.

Initially, the demand drawn from the deck of cards remains fairly stable and the players manage to smoothly operate the supply chain. However, soon they start to experience shortages, which they may try to compensate by ordering high quantities of the product just to find out that the product not only arrived too late not to incur in backlog costs. In order to get rid of the mounting holding inventory costs due to over-ordering, they might try to reduce the order quantities and, but in doing so, they might experience shortages again. After completing the simulation, a debriefing shows the amplification of the orders upstream the supply chain typical of the bullwhip effect.

Throughout the data obtained by players of the beer game, it is now well-established that people use to consistently underestimate the pipeline inventory (i.e. orders placed in the past that have not yet arrived) when making ordering decisions. This has been also observed in empirical data. It is remarkable that this bias has been detected even in experiments where the demand information and the stock position of each member of the SC are known.

3.4 Operational Causes

Are the behavioural causes described in the previous section the only reason influencing the bullwhip effect? If so, then a natural solution for the amplification of the variance of orders and inventories would be given by *rationalising* (or, in other words, *optimising*) the decisions made in managing the SC. Unfortunately, in the 90s, it was found that this was not the case and, indeed, in some cases, it was the *rational* behaviour of the decision makers a factor that may exacerbate the bullwhip effect. Although we will provide a detailed explanation of some of these structural causes of the bullwhip effect throughout this book, it is perhaps interesting to give some intuition about the common pattern of these causes. These are as follows:

- **Demand forecast updating**. Since, in many cases, the future demand is not known, it has to be forecasted using some forecasting techniques (some of these will be discussed in Chap. 2). These techniques usually rely on past demand information, which is continuously updated as new demand data become available over time.

Usually, a surplus amount of product is added on top of the expected demand to deal with demand uncertainty, so this safety stock may be used to cover the unforeseen demand, as we have seen in Chap. 2. Clearly, the longer the time between replenishment, the higher the safety stock required. The readjusting of the demand information causes the safety stocks to be readjusted and this in turn creates a higher fluctuation in the variability of the orders (in some periods, smaller quantities are required as there is a lot of safety stocks, and in some other periods, a high quantity of products is ordered to build up the exhausted safety stock).
Furthermore, this *natural* uncertainty in the demand can be further exacerbated by both structural aspects of the SC (such as the existence of the return of products at the end of its life) or operational aspects (such as the unreliability of the inventory records or the timeliness of the information transmitted to or received from other nodes in the SC). Given the ubiquity of these phenomena, this cause of the bullwhip effect will be discussed extensively in the models presented in the remainder of the book.

- **Order batching**. In many companies, the granularity of the decisions regarding demand fulfilment and inventory replenishment is not the same. That is, whereas demand fulfilment (i.e. matching customer demand with existing inventory) may be done on a daily (or even more frequent) basis, the company may place an order with its supplier once a week or even longer periods of time. This is known as *order batching*, and it is usually caused by economic reasons: on many occasions, the costs and time to process an order can be substantial.[1] In this situation, even a smooth pattern of demand may translate into much more variable order patterns. When a company faces such long period ordering cycles by its customers, the bullwhip effect might be amplified by the joint effect of the ordering from these multiple customers. If the order cycles of several customers coincide (e.g. they all order by the end of each month), then the order variability is augmented. An example of this type of cause will be modelled and discussed in Sect. 8.2.
- **Price fluctuation**. In some sectors (e.g. the grocery industry), a high percentage of the transactions are made in a so-called *forward buy* arrangement (i.e. the items are actually purchased in advance of requirements), usually due to an attractive price offer made by the supplier. When this practice becomes the norm, the customer buys in bigger quantities than needed whenever there is a discount or price offer and avoids buying at a regular price unless its inventory is depleted. As a result, the ordering pattern does not follow the demand pattern, and the variability of the orders is much higher than that of the demand. Note that this behaviour is indeed a very rational decision on the customer side, as usually the inventory holding costs are lower than the difference in prices. However, it is damaging the SC performance by forcing companies to build up higher inventory levels.

[1] For instance, transportation costs are rarely linear: as the prices of Full-Truck Load (FTL) are very different to less-than-truckload, thus favouring order sizes that fill or nearly fill a truck or a container, even if for many products this means ordering once a month or less.

This cause of the bullwhip effect would not be discussed further in the book, as models do not provide much more insight than the intuition explained above. However, we will discuss briefly in Sect. 3.5 the approaches that may be used to mitigate it.

- **Rationing and shortage game**. When product demand exceeds supply, the manufacturer may have to ration its products to the customers. These sudden surges in the demand for a product may be caused to anticipate a perceived or real future scarcity of a product and are relatively common, with many cases documented in the computer and electronics industry. An apparently rational way to perform such rationing is to allocate the available product to each customer according to the ratio between the amount ordered by this customer and the total amount ordered by all customers. This makes sense as all customers are equally treated. However, if this is the policy adopted by the manufacturer, then it is very *rational* for each customer to ask for a higher amount than its real needs in order to increase its allocation ratio. If all customers adopt this (locally optimal) policy, then the total amount of orders received by the manufacturer would be higher than in the previous period. At some point, however, the real or perceived shortage of the product would end, and then the volume of orders would reach their usual values or even less so the customers can absorb the incoming product. Or, even worse, if it is possible for the customers to cancel their orders, they will do so. Clearly, all this process inflates order variability.
  As in the price fluctuation, the insights that can be obtained by models do not add much to the discussion above, and therefore, we would not further elaborate on the models, although we will discuss the ways to tackle these causes.

An important consequence of the findings of these studies is that there are operational factors influencing the bullwhip effect and, since they are structural (as opposed to, e.g. psychological or subjective), they can be explained using formulae from models. In order to provide such an explanation, we first need to provide suitable, objective indicators of the bullwhip effect so they can be incorporated into these models. This is done in the next section.

3.5 Approaches to Mitigate the Bullwhip Effect

The classification of the causes of the bullwhip effect among behavioural and operational may, in principle, establish different fronts to combat the bullwhip effect. Clearly, the potential solutions to the operational causes lay in the companies' ability to modify the typical SC infrastructure and to a lesser extent (but, as we will see, not completely unrelated) to modify decision makers' behaviour. In the next subsections, we discuss these approaches depending on the specific stream.

3.5.1 Behavioural Causes

With respect to the potential mitigation of behavioural causes, it seems that the transmission of the dynamic information across the members of the supply chain helps upstream members to better interpret orders on the part of their customers and to prevent them from overreacting to the fluctuations when placing their own orders. On the other hand, since even trained experts cannot somehow avoid the bias in the estimation of the pipeline inventory, it is perhaps useful to take this aspect into account when assessing the potential advantages of increasing information transparency across the supply chain, as it also may have the side effect of reducing the pipeline bias. Furthermore, the role of training SC managers does not seem to be a silver bullet to mitigate the behavioural causes of the bullwhip effect, as it has been observed that the underweight of the pipeline inventory has been also observed for subjects who are trained logistic professionals, and that risk aversion in replenishment decisions persists even when experience is gained. However, since there are high potential cost savings in removing the behavioural causes of the bullwhip effect, it is clear that the training effort cannot be abandoned.

3.5.2 Operational Causes

Regarding approaches to mitigate the operational causes, some of them seem quite clear as the cause can be isolated. This happens, e.g. with the price fluctuation, whose remedy (at least in theory) is clear: if there is no price fluctuation, no bullwhip effect can be assigned to this cause. In other cases (such as in the case of the demand forecast updating), the situation is much more complex, as there are interdependencies between several decisions such as the forecasting method and the replenishment policy adopted. In this section, we will discuss the remedies for these causes that are more easy to identify and will just briefly outline the approaches in the other cases, which will be better understood when the corresponding models are developed. Table 3.1 shows a summary of the aforementioned practices.

Regarding the price fluctuation, the simplest way to control the bullwhip effect is to reduce both the frequency and the level of the price discount. In practice, this means changing the price strategy from promotions to an Every Day Low Price (EDLP) which, in the long run, seems to bear fruits in terms of profit margins. However, EDLP is far from being a standard practice in many companies and, while it is clear that the promotions would increase the bullwhip effect and its associated costs, there are models explaining why a careful design of the promotion may compensate for such increase. Another way that can potentially reduce the attractiveness of the promotions for a company is to implement a cost accounting system that takes into

Table 3.1 Summary of the operational causes of the bullwhip effect (adapted from [1])

Causes of bullwhip	Information sharing	Channel alignment	Operational efficiency
Forecast demand update	Use POS data	VMI Consumer direct	Lead time reduction
Order batching	EDI	3PL	Reduction in fixed order costs
Price fluctuations			Cost accounting EDLP practices
Shortage gaming	Sharing sale, capacity or inventory data	Allocation of orders based on past sales Penalties for order cancellation	

account the real costs of forwarding buying practices,[2] which may not show up in some accounting systems.

Regarding the shortage and gaming, it seems clear that allocating the product based on the current orders is not adequate in a shortage situation. Instead, it can revert then to allocate the product according to past sales record. By doing so, the customers have no incentives to exaggerate their orders. Another line is to act upon the sometimes perceived situation of scarcity by providing visibility on its inventory (or production capacity) to the customers to remove this perception. Naturally, this would not work if the scarcity is real. A final avenue to disincentive gaming practices is to impose penalties on order cancels. If these penalties are proportional to the number of units ordered (even if they may be extremely low), it could deter some companies to escalate their orders beyond a certain limit.

Regarding order batching, a clear action is to reduce the replenishment intervals, for which the fixed order costs have to be reduced. This can be done within the company via achieving higher operational efficiency or by outsourcing this business function so it can be performed in a more efficient manner. There are two main aspects influencing fixed order costs: order processing costs and transportation costs. In the first case, the deployment of adequate IT infrastructures to perform computer-assisted ordering is known to reduce the order processing costs by saving clerical time. With respect to the transportation costs, perhaps the obvious solution is to outsource it to 3PL (Third-Party Logistics) companies, which can clearly use economies of scale to lower the otherwise highly non-linear transportation costs.[3]

[2] Here, we can mention a practice that can be called *product diversion*, referring to a situation where a company runs a regional promotion, and then the retailers located in this region buy high numbers of the product, not only for their consumption in this region, but also to divert these products for the consumption in other regions where this promotion is not in place.

[3] The differences in the cost of a full truckload and that of a less-than-truckload are usually very high, making the unit transportation costs to be non-linear.

Regarding the demand forecast updates, the most effective response so far seems to be to make demand data available to each node in the supply chain (possibly by sharing Point-of-Sale or POS data) in order to avoid multiple demand forecast updates. In many cases, the technological infrastructure to share data already exists (Electronic Data Interchange or EDI is widely used for the transmission of orders).

However, even if all the nodes in the SC have the same demand data, the differences in forecasting techniques may lead to fluctuations in the orders placed to the upstream node. An approach to remove the need for multiple forecasts is that the upstream node of the SC performs the forecast and replenishment operations for the downstream node (who in turn does not have to perform such operations and its tasks are limited to facilitating the demand data to the upstream node). This is the idea behind common practices in some sectors known as Vendor-Managed Inventory (VMI), Continuous Replenishment Program (CRP) or CPFR (Continuous Planning Forecasting and Replenishment).

In addition, the need for several nodes in the SC has to be questioned in terms of their contribution to add value for the final customer. If this is not the case, the aforementioned problems may lead to a change in the role of some nodes in the SC in order to increase their visibility over the final demand. Perhaps the most employed initiative in this regard is the so-called *consumer direct* programs, where the manufacturer sells directly to the final customer bypassing resellers and distributors.

Finally, it is clear that the bullwhip effect increases if the lead times are longer, as this creates the need for the node to face the existing inventory with the unknown demand during a higher number of periods (recall Eq. (2.55) in Sect. 2.4); therefore, another avenue is to increase the operational efficiency of the SC by reducing the lead times.

3.6 Performance Measures in SC/SCD

After discussing the two streams of causes in the bullwhip effect, maybe, it is tempting to state that structural causes are more difficult to handle than behavioural ones, given the fact that the former are inherent to the SC design and because we naturally tend to be optimistic about the possibility that the organisations change its behaviour. However, this is not the case in many organisations where there is a lot of inertia. In any case, it seems that none of the possible actions discussed in Sect. 3.5 is free, as they would require, in some cases, substantial investments to be implemented. Therefore, it is clear that we need to measure the damaging effect of the bullwhip effect in order to assess whether the benefits of these actions outperform the investment required.

Perhaps an initial question when presenting indicators of the bullwhip effect is precisely to discuss if these indicators are really required. After all, we can assess in a relatively easy manner if certain inventory management or order replenishment policy in a company is beneficial or not, just by evaluating if such policy positively impacts the bottom line of the company or not. This is certainly not different in the case of the bullwhip effect since—as we already discussed in Chap. 1—the amplification in the inventory variance causes extra inventory costs as losses in lost customers due to a poor service level. In this regard, it sounds legitimate to ask if it is then sufficient to use the benefits (costs) as an indicator of the bullwhip effect.

Although it is clear that financial measures are the ultimate indicator of the bullwhip effect, these measures are closely tightened to the cost structure of the companies and the SCs, and therefore do not easily allow to isolate the operational effects from other operational factors. For instance, in a company in which its holding inventory costs are remarkably low, it may be that the bullwhip effect goes almost unnoticed, at least in terms of inventory variability. However, these low costs may be caused by an excellent performance of the company or because these costs for all companies in the sector are low. We see that it is not possible to differentiate these just based on financial indicators. Furthermore, such indicators do not allow for easy comparison among nodes within the same SC or among different SCs.

Therefore, it seems more interesting to develop some indicators that are cost-independent. After all, these indicators can be later be transformed easily into cost—something that is really useful if we want to measure the economic impact of the bullwhip effect—but they allow for an easier comparison of scenarios.

In order to measure the dynamic performance of a supply chain, there are two criteria that can be used:

- Scope: Local versus global measures. Local indicators serve to assess the dynamic performance of a single node in the supply chain or the dynamic performance of the supply chain as a whole.
- Focus: Internal efficiency versus customer satisfaction. Some indicators (such as the zero replenishment discussed in Sect. 3.6.4) are oriented towards the internal efficiency of the operations of the node, whereas other indicators (such as the backlog or fill rate, also discussed in Sect. 3.6.4) are more suitable to measure the customer satisfaction.

Although in this book, we will use primarily two indicators of internal efficiency at the local level (BWE and $NSAmp$, discussed in Sects. 3.6.1 and 3.6.2), we will also discuss most of the indicators employed in the literature, as well as their relationship among them.

3.6.1 Measuring Order Variability

Since we are interested in measuring the amplification of the variability of certain magnitudes across the SC, it is natural to use ratios between the variances. The most widely known measure among this type of indicator is BWE, defined as follows:

$$BWE = \frac{V[\text{Orders}]}{V[\text{Demand}]} \tag{3.1}$$

where V[Orders] or $V[O]$ is the variance of the orders issued by the node to its provider. Similarly, V[Demand] is the variance of its customer demand. If the indicator is employed to measure an specific echelon i in the supply chain, then it can be written as follows:

$$BWE_i = \frac{V[O_i]}{V[D_i]} \tag{3.2}$$

Note that, for any node different than the node 1, the variance of the customer demand is the variance of the orders issued by the customer, so the above expression can be written as (recall the convention to number the echelons presented in Chap. 1) follows:

$$BWE_i = \frac{V[O_i]}{V[O_{i-1}]} \tag{3.3}$$

with $V[O_0] = \sigma^2$ for consistency reasons.

Another related measure is the so-called *Order Variance Ratio* or ORV of a node in the SC. ORV for an echelon i is defined as its order variance divided by the variance of the final customer demand:

$$ORV_i = \frac{V[O_i]}{V[\sigma^2]} \tag{3.4}$$

Clearly, both indicators are related and it is easy to show that the following equivalence holds:

$$ORV_i = \Pi_{k=1}^{i} BWE_k \tag{3.5}$$

For both indicators, a value larger than one indicates that the bullwhip effect is present (amplification), whereas values smaller than one refer to a scenario where the output variability is smaller than the input variability (dampening). This metric provides information on the potential extra costs incurred by an echelon in the SC since usually a node uses a safety stock to protect against the variability of its demand (see Sect. 2.4 in Chap. 2) and therefore, it incurs in additional costs caused by this amplification of the variability.

Note that, in practice, it is easier for a company to measure its BWE rather than its ORV since, on most occasions, information regarding the demand of the final customer is not available for nodes in echelons 2, 3, etc. And, since we have shown that they are related, we will focus our analysis in the next chapters in BWE.

Finally, there are additional indicators of the amplification of the variance that have been proposed in the literature. Perhaps the best known is the so-called Order Rate Variance Ratio (ORVR), which essentially scales the amplification of the variance to its mean value, i.e.

$$ORVR_i = \frac{\frac{V[O_i]}{E[O_i]}}{\frac{\sigma^2}{\mu}} \tag{3.6}$$

Note however that, in many situations, it is expected that the long-run average of the orders and that of the demand are the same, so this indicator basically reduces to the ORV unless some phenomena (e.g. such as some versions of the IRI discussed in Chap. 5) provoke a distortion of the average orders across the SC.

3.6.2 Measuring Inventory Variability

The variance in the demand coupled with the ordering policies would cause fluctuations in the inventory levels of each node in the SC. Similarly to with BWE_i, the following indicator—denoted as $NSAmp_i$ or Net Stock Amplification—can be used to measure such fluctuations:

$$NSAmp_i = \frac{V[I_i]}{V[D_i]} \tag{3.7}$$

where $V[I_i]$ denotes the variance of inventory level of echelon i and $V[D_i]$ the variance of the demand perceived by the echelon.

Measuring the variance of the inventory is important since the higher the variance of the inventory, the higher the safety stock required to attain a given service level. Clearly, such safety stock is also influenced by the variability of the demand (the more variable the demand, the higher the safety stock required); therefore, the indicator is measuring the potential extra costs incurred once the effect of the demand variability has been discounted.

In the remainder of the book, we will use sometimes $NSAmp$ as an indicator, although as we will see, in some cases, it is quite complex to obtain a closed expression and we will stick only to BWE as an indicator of the SC dynamics.

3.6.3 Slope Metrics

Clearly, the indicators presented before are not appropriate to compare the global performance of different SCs, so it will be interesting to condensate the performance of the whole SC along one or several indicators that can be used to assess one aspect (such as, e.g. BWE, $NSAmp$ or other indicators) of the dynamics of the SC. One idea to visualise the overall effect of, e.g. the order variability amplification BWE would be to plot, for each echelon of the supply chain, its corresponding BWE value. Typically, we would obtain an increasing curve such as the one shown in Fig. 3.1. Qualitatively, if there is no big amplification of the order variance, one would expect something similar to a linear trend in this plot, whereas a higher-than-linear trend would indicate an exacerbation of the amplification across the SC. We can use this idea to develop a metric that allows us to compare two SCs according to a given indicator: for each echelon i in each SC, we can first plot the corresponding local

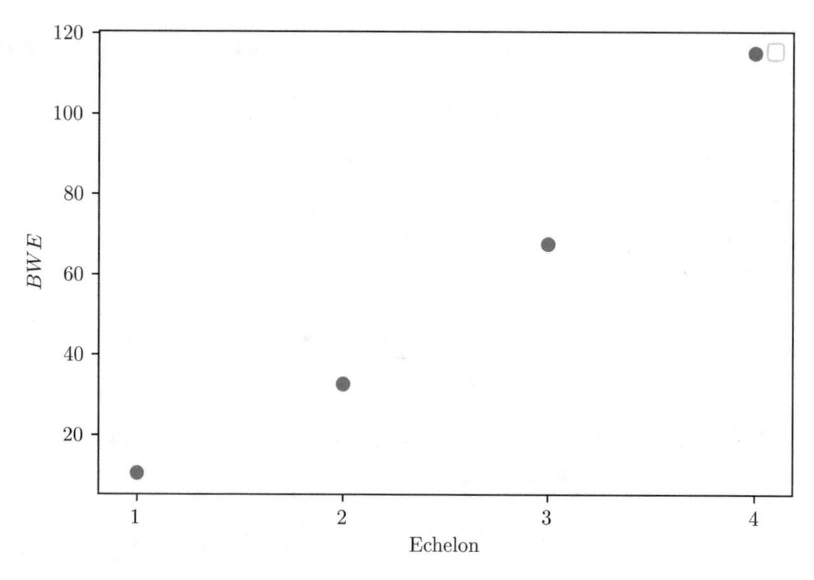

Fig. 3.1 Typical plotted values of BWE for a 4-echelon SC

indicator of the bullwhip effect M_i and find the most appropriate regression model for each one of the SC. Clearly, the order of the best-fitting regression model would indicate the growth ratio of the indicator. If the best-fitting model for both SCs under comparison is the same, then we can compute the slope coefficient and check in which of these two SCs the growth is higher. Let us see next some examples in the case that the plot is best represented by a linear regression model.

If we assume that there is a linear relationship between the independent variable x and the dependent variable y, i.e. their relationship follows the model

$$y = \alpha + \beta \cdot x \tag{3.8}$$

Then, if we have n samples (x_i, y_i) with $i = 1, \ldots, n$ it is known that the slope β can be estimated using the following formula (let us denote by $\hat{\beta}$ such estimation):

$$\hat{\beta} = \frac{\sum_{i=1}^{n}(x_i - \bar{x})(y_i - \bar{y})}{\sum_{i=1}^{n}(x_i - \bar{x})^2} \tag{3.9}$$

which, after an easy algebraic manipulation, can be rearranged as

$$\hat{\beta} = \frac{n \cdot \sum_{i=1}^{n} x_i \cdot y_i - \left(\sum_{i=1}^{n} x_i\right)\left(\sum_{i=1}^{n} y_i\right)}{n \cdot \sum_{i=1}^{n} x_i^2 - \left(\sum_{i=1}^{n} x_i\right)^2} \tag{3.10}$$

Therefore, if we assume that there is a linear relationship between p_i the position of each node i in the SC (i.e. the echelon where the node i belongs to) and its corresponding (local) metric of the bullwhip effect M_i, then the slope of such model would be

$$\hat{\beta} = \frac{n \cdot \sum_{i=1}^{n} p_i \cdot M_i - \left(\sum_{i=1}^{n} p_i\right)\left(\sum_{i=1}^{n} M_i\right)}{n \cdot \sum_{i=1}^{n} p_i^2 - \left(\sum_{i=1}^{n} p_i\right)^2} \tag{3.11}$$

This slope indicator can be applied, in principle, to any local metric of the bullwhip effect. Therefore, for instance, one can define the BWE slope or $BWESl$ in the following manner:

$$BWESl = \frac{n \cdot \sum_{i=1}^{n} p_i \cdot BWE_i - \left(\sum_{i=1}^{n} p_i\right)\left(\sum_{i=1}^{n} BWE_i\right)}{n \cdot \sum_{i=1}^{n} p_i^2 - \left(\sum_{i=1}^{n} p_i\right)^2} \tag{3.12}$$

Similarly, $NSAmpSl$ the slope of the amplification of the inventory variance across the supply chain can be defined as

$$NSAmpSl = \frac{n \cdot \sum_{i=1}^{n} p_i \cdot NSAmp_i - \left(\sum_{i=1}^{n} p_i\right)\left(\sum_{i=1}^{n} NSAmp_i\right)}{n \cdot \sum_{i=1}^{n} p_i^2 - \left(\sum_{i=1}^{n} p_i\right)^2} \tag{3.13}$$

As we can see, the idea of the slope can be used to compare the performance of several SCs at a global level. However, since most of the models presented in this book would refer to the dynamics experienced between a node and its provider, these global measures would not be used extensively.

3.6.4 Other Metrics

In addition to the metrics exposed in the previous sections, there are additional indicators that can be employed to measure the dynamics of the SC, both at a local and a global level.

One metric that is connected to the internal efficiency is the so-called *zero-replenishment* indicator, which measures the number of times that, during a given time horizon, a node does not issue orders to its supplier. This usually occurs when the number of units in the inventory largely exceeds the target inventory level. Therefore, an order pattern characterised by a significant number of periods with zero replenishment is indicative of an over-dimensioning of the previous orders, and therefore, it may point out to an inefficient replenishment policy or to an unsuitable demand forecast, both factors related to the internal efficiency of the node.

Another, more conventional indicator is to measure the backlog or the fill rate. Backlog is a cumulative measure that is more appropriate to monitor a build-to-order system as it provides information on the conformance to specification, whereas fill rate is a measure that is more suitable to assess the percentage of lost sales in the case of mass products and make-to-stock systems.

3.7 Summary

In companies, both behavioural and operational causes of the bullwhip effect do co-exist. Indeed, it has been shown that the bullwhip effect exists if one of the causes is removed, but the other persists. It has also been shown that some techniques such as information sharing may attack both causes.

Since addressing most of the causes of the bullwhip effect requires some investment, it is necessary to adequately measure the bullwhip effect in order to assess the efficiency of the countermeasures to be implemented. In this regard, we have extensively discussed in this chapter a number of indicators and metrics that can be used (alone or in combination) to evaluate the bullwhip effect both at a single-node (local) or SC (global) levels.

From this chapter, we can see that within the structural causes, some of them refer to the interaction of different operational decisions and that the effects can be mitigated or exacerbated both by structural and operational factors. It is, therefore, necessary to understand the combined result of these factors, for which suitable models have to be developed. This will be the goal of the next chapter.

3.8 Further Readings

A recent review of the literature on the behavioural causes of the bullwhip effect is [2]. The empirical evidence of the people underestimating the pipeline inventory is from [3]. The experiments showing the prevalence of the behavioural causes of the bullwhip effect even if the demand is known are due to [4]. An study on risk aversion in replenishment decisions is [5]. In [6], an experiment showing that the SC managers do not use all available information is presented. Regarding the potential cost saving by addressing behavioural causes, [7], based on a simulation of the Beer Game scenario, states that removing the bullwhip effect may reduce SC costs by 50%. The need for investments in training programs in order to increase managers' ability to perceive, understand and properly react to the dynamics of the SC is first discussed in [8]. A number of references (including [9–12]) discuss the persistence of the behavioural causes even among SC managers, a fact stressed by the study by [13] showing that decision makers create bullwhip even under the most information-rich conditions. These difficulties to address the behavioural causes of the bullwhip effect have been explained using different experimental frameworks, including [14] using linear control theory or [15] in terms of analogical reasoning.

Regarding the operational causes of the bullwhip effect, here the seminal references are [1, 16]. Note that, although the four operational causes described in Sect. 3.4 have become the standard to interpret the bullwhip effect, there are alternative proposals to classify the causes, such as [17] or [18]. Some of these additional causes for the bullwhip effect were early mentioned by [19]. Also note that, although [1, 16] present a coherent and unifying picture of the bullwhip effect, some of its causes have been already discussed by some scholars, who show that the bullwhip effect may stem from rational actions of stakeholders trying to avoid stockouts [20], to reduce production costs [21] or to respect to unexpected demand shocks [22]. In [23], the impact of price promotions and other marketing initiatives on the SC dynamics is studied. In the references [24] or [25], the reader might find several documented cases of the shortage game discussed in Sect. 3.4. An in-depth analysis of the rationing game in SCs can be found in [26], where it is shown that it is possible that there are bounds to the increasingly inflated orders placed by the retailers. A further reference studying the bullwhip effect under the rationing game is [27]. With respect to the impact of price fluctuations in the SC dynamics, some cases can be found in [28, 29].

Regarding the approaches to mitigate the bullwhip effect, price stabilisation initiatives are discussed in [30, 31], while a detailed discussion on the advantages and disadvantages of EDLP as compared to promotions can be found, e.g. in [32], [33] or [34]. The use of AI methods to make optimal replenishment decisions aimed at reducing the bullwhip effect is presented in [35]. In [36], the benefits of VMI in terms of bullwhip reduction are investigated in a field study, whereas in [37], an adaptive VMI is proposed to reduce the bullwhip effect to a greater extent than in the traditional VMI. A discussion on the merits of CPFR for smoothing the SC dynamics can be found in [38], whereas the effect on the bullwhip of another centralised planning initiative (Distribution Resource Planning or DRP) is studied in [39]. An empirical analysis on how ICT can mitigate the bullwhip effect is conducted in [40], and the use of big data and its influence on the bullwhip effect is investigated in [41].

The consideration of the bullwhip effect as a symptom of a poorly performing SC can be found in [42], whereas the idea of using it as a surrogate measure of production adaptation costs is given in [43]. A detailed discussion on the metrics that can be used to assess SC dynamics is given in [44]. The idea of the slope metrics appeared first in [45]. Some additional metrics for the bullwhip effect—some of them able to separate its different causes—have been developed. In this regard, see, e.g. [46–49] or [50]. In [51], a time-variant measure of the bullwhip effect is proposed. A further discussion on the suitability of the different metrics is presented in [52], whereas [53] studies the relation between order variability amplification and inventory variability amplification. The challenges of estimating and measuring the bullwhip effect using empirical data have been discussed in [54].

References

1. Lee, H., Padmanabhan, V., Whang, S.: The bullwhip effect in supply chains. Sloan Manag. Rev. **38**(3), 93–102 (1997)
2. Yang, Y., Lin, J., Liu, G., Zhou, L.: The behavioural causes of bullwhip effect in supply chains: a systematic literature review. Int. J. Prod. Econ. **236**, 108,120 (2021)
3. Sterman, J.D.: Modeling managerial behavior: misperceptions of feedback in a dynamic decision making experiment. Manag. Sci. **35**(3), 321–339 (1989)

4. Croson, R., Donohue, K.: Behavioral causes of the bullwhip effect and the observed value of inventory information. Manag. Sci. **52**(3), 323–336 (2006)
5. Di Mauro, C., Ancarani, A., Schupp, F., Crocco, G.: Risk aversion in the supply chain: evidence from replenishment decisions. J. Purch. Supply Manag. **26**(4) (2020)
6. Haines, R., Hough, J., Haines, D.: Individual and environmental impacts on supply chain inventory management: an experimental investigation of information availability and procedural rationality. J. Bus. Logist. **31**(2), 111–128 (2010)
7. Rinks, D.: System dynamics in supply chains. In: Proceedings of the 2002 Euroma Conference, pp. 443–457 (2002)
8. Senge, P., Sterman, J.: Systems thinking and organizational learning: acting locally and thinking globally in the organization of the future. Eur. J. Oper. Res. **59**(1), 137–150 (1992)
9. Tokar, T., Aloysius, J., Waller, M.: Supply chain inventory replenishment: the debiasing effect of declarative knowledge. Decis. Sci. **43**(3), 525–546 (2012)
10. Ancarani, A., Di Mauro, C., D'Urso, D.: Measuring overconfidence in inventory management decisions. J. Purch. Supply Manag. **22**(3), 171–180 (2016)
11. Tokar, T., Aloysius, J., Waller, M., Hawkins, D.: Exploring framing effects in inventory control decisions: violations of procedure invariance. Prod. Oper. Manag. **25**(2), 306–329 (2016)
12. Turner, B., Goodman, M., Machen, R., Mathis, C., Rhoades, R., Dunn, B.: Results of beer game trials played by natural resource managers versus students: does age influence ordering decisions? Systems **8**(4), 1–30 (2020)
13. Haines, R., Hough, J., Haines, D.: A metacognitive perspective on decision making in supply chains: revisiting the behavioral causes of the bullwhip effect. Int. J. Prod. Econ. **184**, 7–20 (2017)
14. Udenio, M., Vatamidou, E., Fransoo, J., Dellaert, N.: Behavioral causes of the bullwhip effect: an analysis using linear control theory. IISE Trans. **49**(10), 980–1000 (2017)
15. Naim, M., Spiegler, V., Wikner, J., Towill, D.: Identifying the causes of the bullwhip effect by exploiting control block diagram manipulation with analogical reasoning. Eur. J. Oper. Res. **263**(1), 240–246 (2017)
16. Lee, H., Padmanabhan, V., Whang, S.: Information distortion in a supply chain: the bullwhip effect. Manag. Sci. **43**(4), 546–558 (1997)
17. Miragliotta, G.: Layers and mechanisms: a new taxonomy for the bullwhip effect. Int. J. Prod. Econ. **104**(2), 365–381 (2006)
18. Bhattacharya, R., Bandyopadhyay, S.: A review of the causes of bullwhip effect in a supply chain. Int. J. Adv. Manuf. Technol. **54**(9–12), 1245–1261 (2011)
19. Taylor, D.: Measurement and analysis of demand amplification across the supply chain. Int. J. Logist. Manag. **10**(2), 55–70 (1999)
20. Kahn, J.: Inventories and the volatility of production. Amer. Econ. Rev. **77**(4), 667–679 (1987)
21. Eichenbaum, M.: Some empirical evidence on the production level and production cost smoothing models of inventory investment. Amer. Econ. Rev. **79**(4), 853–864 (1989)
22. Naish, H.: Production smoothing in the linear quadratic inventory model. Quart. J. Econ. **104**(425), 864–875 (1994)
23. Lummus, R., Duclos, L., Vokurka, R.: The impact of marketing initiatives on the supply chain. Surg. Endosc. Other Interv. Tech. **8**(4), 317–323 (2003)
24. Lode, L.: The role of inventory in delivery time competition. Manag. Sci. **38**(2), 182–197 (1992)
25. Kelly, K.: Burned by busy signals: Why motorola ramped up production way past demand. Business Week (6), 36 (1995)
26. Rong, Y., Snyder, L., Shen, Z.J.: Bullwhip and reverse bullwhip effects under the rationing game. Nav. Res. Logist. **64**(3), 203–216 (2017)
27. Bray, R., Yao, Y., Duan, Y., Huo, J.: Ration gaming and the bullwhip effect. Oper. Res. **67**(2), 453–467 (2019)
28. Buzzell, R.D., J.Q., Salmon, W.: The costly bargain of trade promotions. Harvard Business Review pp. 141–148 (1990)
29. Sellers, P.: The dumbest marketing ploy. Fortune **126**(5), 88–93 (1992)

30. Schiller, Z.: Ed artzt's elbow grease has pg shining. Bus. Week **10**, 84–86 (1994)
31. Mathews, R.: Crp moves towards reality. Progress. Groc. **73**, 43–44 (1994)
32. Su, Y., Geunes, J.: Price promotions, operations cost, and profit in a two-stage supply chain. Omega **40**(6), 891–905 (2012). Special Issue on Forecasting in Management Science
33. Su, Y., Geunes, J.: Multi-period price promotions in a single-supplier, multi-retailer supply chain under asymmetric demand information. Ann. Oper. Res. **211**(1), 447–472 (2013)
34. Park, I., Jung, I., Choi, J.: Market competition and pricing strategies in retail supply chains. Manag. Decis. Econ. **41**(8), 1528–1538 (2020)
35. O'Donnell, T., Humphreys, P., McIvor, R., Maguire, L.: Reducing the negative effects of sales promotions in supply chains using genetic algorithms. Expert Syst. Appl. **36**(4), 7827–7837 (2009)
36. Dong, Y., Dresner, M., Yao, Y.: Beyond information sharing: An empirical analysis of vendor-managed inventory. Production and Operations Management **23**(5) (2014)
37. Kristianto, Y., Helo, P., Jiao, J., Sandhu, M.: Adaptive fuzzy vendor managed inventory control for mitigating the bullwhip effect in supply chains. Eur. J. Oper. Res. **216**(2), 346–355 (2012)
38. Holmström, J., Främling, K., Kaipia, R., Saranen, J.: Collaborative planning forecasting and replenishment: new solutions needed for mass collaboration. Surg. Endosc. Other Interv. Tech. **7**(3), 136–145 (2002)
39. Nguyen, D., Adulyasak, Y., Landry, S.: Research manuscript: the bullwhip effect in rule-based supply chain planning systems-a case-based simulation at a hard goods retailer. Omega (United Kingdom), vol. 98 (2021)
40. Yao, Y., Zhu, K.: Do electronic linkages reduce the bullwhip effect? an empirical analysis of the u.s. manufacturing supply chains. Inf. Syst. Res. **23**(3 PART 2), 1042–1055 (2012)
41. Hofmann, E.: Big data and supply chain decisions: the impact of volume, variety and velocity properties on the bullwhip effect. Int. J. Prod. Res. **55**(17), 5108–5126 (2017)
42. Jones, D., Simons, D.: Future directions for the supply side of ecr. ECR in the Third Millennium - Academic Perspectives on the Future of Consumer Goods Industry pp. 34–40 (2000)
43. Stalk, G., Hout, T.: Competing Against Time (1990)
44. Cannella, S., Barbosa-Póvoa, A., Framinan, J., Relvas, S.: Metrics for bullwhip effect analysis. J. Oper. Res. Soc. **64**(1), 1–16 (2013)
45. Dejonckheere, J., Disney, S., Lambrecht, M., Towill, D.: The impact of information enrichment on the bullwhip effect in supply chains: a control engineering perspective. Eur. J. Oper. Res. **153**(3 SPEC. ISS.), 727–750 (2003)
46. El-Beheiry M., W.C.Y.E.K.A.: Empirical quantification of the bullwhip effect. In: Proceedings of the 13th Working Seminar on Production Economics, vol. 3, pp. 259–274 (2004)
47. Metters, R.: Quantifying the bullwhip effect in supply chains. J. Oper. Manag. **15**(2), 89–100 (1997)
48. Fransoo, J., Wouters, M.: Measuring the bullwhip effect in the supply chain. Surg. Endosc. Other Interv. Tech. **5**(2), 78–89 (2000)
49. Riddalls, C., Bennett, S.: The optimal control of batched production and its effect on demand amplification. Int. J. Prod. Econ. **72**(2), 159–168 (2001)
50. Disney, S., Towill, D.: The effect of vendor managed inventory (vmi) dynamics on the bullwhip effect in supply chains. Int. J. Prod. Econ. **85**(2), 199–215 (2003)
51. Trapero, J., Pedregal, D.: A novel time-varying bullwhip effect metric: An application to promotional sales. Int. J. Prod. Econ. **182**, 465–471 (2016)
52. Vicente, J., Relvas, S., Barbosa-Póvoa, A.: Effective bullwhip metrics for multi-echelon distribution systems under order batching policies with cyclic demand. Int. J. Prod. Res. **56**(4), 1593–1619 (2018)
53. Chen, L., Luo, W., Shang, K.: Measuring the bullwhip effect: Discrepancy and alignment between information and material flows. Manuf. Serv. Oper. Manag. **19**(1), 36–51 (2017)
54. Yao, Y., Duan, Y., Huo, J.: On empirically estimating bullwhip effects: Measurement, aggregation, and impact. J. Oper. Manag. **67**(1), 5–30 (2021)

Chapter 4
Basic Models for SC Dynamics

4.1 Introduction

In this chapter, a basic model to understand the dynamic behaviour of the supply chain is presented. First, a brief introduction on the role of modelling in the context of Operations Management/Supply Chain Management is given, so the reader can understand the process of abstraction of the essential hypotheses and constraints from a real-world problem, the formulation of the model and the extrapolation of the analysis of the model to the real-world problem. Then, the main notation, equations and sequences of events in a basic, two-level supply chain are presented in great detail: a node from the supply chain is conceptualised and a notation is given for the material and information flows, and the main hypotheses and assumptions of the basic model are discussed. Next, a two-echelon basic model is developed so a closed formula is obtained for the main indicator of the bullwhip effect. This formula will allow us to discuss the implications of the model, so the causes and remedies of the bullwhip effect presented in Chap. 3 are revisited in light of the model. Finally, the insertion of the equations at node level into a serial supply chain is presented, along with a discussion on the managerial implications.

More specifically, in this chapter, we:

- Discuss the role of modelling in Operations Management and Supply Chain Management (Sect. 4.2).
- Describe the basic elements of a two-echelon SC that we will consider when developing the most basic model to understand SC dynamics (Sect. 4.3), as well as the indicators to measure it (Sect. 4.4).
- Describe the sequence of events taking place in the model and their translation into a set of equations (Sect. 4.5).
- Discuss the results obtained from the resulting model, together with its extension to an n-echelon SC (Sects. 4.7 and 4.6).

J. M. Framinan, *Modelling Supply Chain Dynamics*,
https://doi.org/10.1007/978-3-030-79189-6_4

4.2 The Role of Modelling in OM/SCM

It is customary to distinguish between analytical and simulation-based SC models. Analytical models describe the behaviour of the system using a number of formulae, so the variables of interest in the system may be expressed as a closed function of a series of input parameters. Clearly, this type of model allows quantifying the effect of these input parameters in the variables of interest (generally by simple inspection), as well as the possibility of deriving the values of these parameters to optimise the variables of interest. Their main drawback is precisely the difficulty to obtain these closed formulae, which sometimes require the formulation of simplifying hypotheses that may restrict their range of application.

On the other hand, simulation models basically replicate in the computer the sequence of events followed by the real system that it is intended to model, either as a whole (discrete-event simulation or continuous models) or the behaviour of each one of the individuals or *agents* in the system, as well as their interactions (agent-based simulation models). Simulation models can be enormously rich as they can capture very complex behaviours, but they are also expensive (in time and resources) to develop and to run. In addition, they are not, in principle, well suited for optimising the variables of interest, and most of the times, this is done basically by some trial and error.

In this book, we will adopt a rather hybrid approach to model the SC. We will start by describing the SC in terms of closed formulae to understand the logic and try to stick to relatively simple hypotheses to do so. Whenever the need for more realistic hypotheses would render impossible (or extremely difficult) to obtain a closed formula of the variables of interest, we will resort to simulate the performance of the system, even if in this case we must not expect the same degree of analysis that we can perform using analytical models.

We believe that this approach has several advantages:

- Rather often, simulation models simply tend to replicate the behaviour of an observed system without distinguishing between specific factors and factors that can be common to other SC. This makes it quite difficult sometimes to extract generalisable conclusions from simulation models.
- The usage of formulae to describe (whenever possible) the behaviour of the system would allow to use relatively simple simulations, which in many cases may be carried out using spreadsheets or a general-purpose programming language.

4.3 Elements of a Basic Supply Chain Model

In our first basic model, we consider a SC of a single product. In this SC model, three elements are considered:

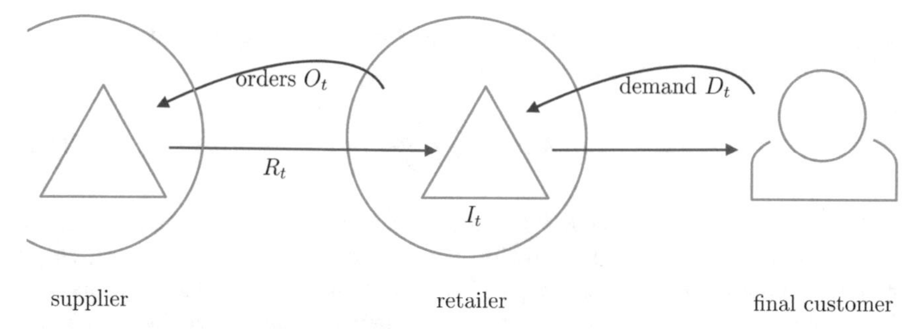

Fig. 4.1 Basic, two-echelon, SC model

- A supplier.
- A retailer.
- The final customer.

As it can be seen in Fig. 4.1, this model represents a two-echelon supply chain (supplier and retailer). The retailer uses her inventory to satisfy the orders (demand) coming from the final customer, that is, she operates in a make-to-stock environment. If the demand cannot be met from inventory, the customer is willing to wait, which generates a backlog of pending orders. To replenish her inventory in order to satisfy future demands from the final customer (or to reduce backlog), the retailer places an order to the supplier. For now, we assume that the supplier always has enough product to meet the retailer's orders; therefore, no backlog is generated in the supplier. Furthermore, the product ordered by the retailer is not immediately available, as the supplier requires some time to produce it and/or to transport it to the retailer. For the time being, this time (*lead time*) is considered as a constant L. As a result of the above hypotheses, the retailer receives in the current period what she has ordered L time periods before.

Even if the model presented is extremely simple, it contains the main elements that are required to understand the SC dynamics: there is a need to anticipate the demand of the final customer (which leads to the need of performing a forecast and to issue orders to face a future demand), and there is a time lag (the lead time) between the period in which the order is placed by the retailer and the period in which the order arrives at its inventory. As the reader would have noticed, the interaction between these three elements in the model can be replicated upstream the SC, since the retailer is the supplier's customer.

4.4 Measuring the Bullwhip Effect

In Sect. 3.6, in Chap. 3, we discussed several metrics to evaluate the bullwhip effect in a SC at a node and a global levels. For the simple two-echelon model that is being described in this chapter, we will simply measure the ratio of the variance of the orders and the variance of the demand, i.e.

$$BWE = \frac{V[\text{Orders}]}{V[\text{Demand}]} \tag{4.1}$$

As already discussed, the BWE indicator in Eq. (4.1) is, without doubt, the most employed measure of the bullwhip effect. Recall that a value of BWE larger than one indicates that the bullwhip effect is present (there is an amplification of the variability), while a value smaller than one indicates that the bullwhip effect is mitigated (or dampened). The value of this indicator has a reflect in the operating costs, as a higher BWE value indicates more frequent changes in the production level, thus meaning higher average product costs per period.

In the basic model to be developed in this chapter, we will also use the $NSAmp$ indicator, defined as

$$NSAmp = \frac{V[\text{Net Stock}]}{V[\text{Demand}]} \tag{4.2}$$

Note that the companies pile up stock to face their demand; therefore, ideally, the pattern of the net stock and that of the demand would have to be the same. The misalignment of these two magnitudes represents an increase in the backlog costs (when demand exceeds net stock) and in the inventory holding costs (when net stock exceeds demand).

4.5 Putting All Pieces Together

To model the dynamics of the SC model described in Sect. 4.3, the time is discretised. The discretisation of the time into single slots forces us to define an (artificial) sequence of events that, in reality, occurs over the continuous time but that we need to sort in order to make the equations consistent. Thus, the chosen sequence of actions for a given period is as follows:

1. The retailer receives from the supplier the orders placed $t - L$ periods before (thus increasing the inventory and decreasing the work in process).
2. The retailer receives the final customer demand and tries to satisfy it with the inventory at hand, thus decreasing the inventory.
3. The retailer studies her needs to face the future customer demand and consequently, she places an order to the supplier, thus increasing the work in process.

Note that this order is purely arbitrary, and a different sequence of events could be devised. This would not affect the conclusions of the models, but will possibly displace the time indices of some variables in the model.

To model the sequence of events, for each time period t, the following variables have to be employed:

- O_t is the order placed by the retailer to the supplier in time period t (i.e. according to the sequence of events, at the end of time period t).
- R_t is the replenishment taking place in time period t, i.e. the number of products received by the supplier in time period t, which correspond to the orders placed by the supplier $t - L$ time periods before. Therefore, the following equation holds:

$$R_t = O_{t-L} \tag{4.3}$$

- D_t is the demand placed to the retailer by the final customer in time period t. In our model, we assume that this demand is satisfied at the beginning of period t *after* receiving R_t; therefore, the products just arrived from the supplier can be used to satisfy the final customer demand (see the sequence of events).
- W_t is the work in process in time period t. The work in process is constituted by the orders that have been placed by the retailer to the supplier but they have not arrived so far. Since we assume that L is the (constant) time from the time period when an order is placed to the time period when the order arrives to the retailer, then it is clear that

$$W_t = \sum_{i=1}^{L-1} O_{t-i} \tag{4.4}$$

In terms of modelling, the sequence of actions discussed before can be represented by different equations using the variables described. This is carried out in the next sections.

4.5.1 Inventory Update

The inventory at the end of time period t has been modified (with respect to the inventory at the beginning of time period $t - 1$) by two events: (1) the arrival of R_t product units and (2) the withdrawal of product units from the inventory to satisfy the demand. Therefore, the following equation holds:

$$I_t = \min(I_{t-1} + R_t - D_t; 0) \tag{4.5}$$

Equation (4.5) expresses the difference of the current inventory at the end of time period t with respect to that of the end of time period $t - 1$ (adding the arrived products from the supplier, subtracting the products sent to the final customer), taking into account that it is not possible for the inventory to take negative values. Nevertheless, this expression introduces a non-linearity in the model that—for now—we would remove by interpreting I_t in a slightly different manner: if non-negative,

I_t represents the inventory level at the end of time period t; however, if negative, it is the backlog generated at the end of time period t. Note that it is not possible that, for a time period t, both the inventory at the end of the period *and* the backlog at the end of the period are non-zero.

With this interpretation, Eq. (4.5) can be rewritten as follows:

$$I_t = I_{t-1} + R_t - D_t \tag{4.6}$$

Although the change may seem of little relevance (and, in some aspects, this is the case), it has some implications when, e.g. computing costs, as the unit inventory costs, are, in general, different than backlog costs; therefore, this distinction has to be taken into account. This aspect would be discussed later on in the book.

It would be useful to write Eq. (4.6) as the difference of inventories at the end of two consecutive time periods, i.e.

$$I_t - I_{t-1} = R_t - D_t \tag{4.7}$$

4.5.2 Work in process Update

The work in process W_t has to be updated accordingly as well. Note that, since the last update of the work in process (which took place in time period $t-1$), two events have modified the value of the work in process: 1) the arrival of the ordered products R_t, which reduces the work in process and 2) the placement of the order O_{t-1} at the end of time period $t-1$—which increases the work in process. Therefore, the following equation holds:

$$W_t = W_{t-1} - R_t + O_{t-1} \tag{4.8}$$

As with the inventory, it is useful to write Eq. (4.8) as the difference of work in process at the end of two consecutive time periods:

$$W_t - W_{t-1} = O_{t-1} - R_t \tag{4.9}$$

4.5.3 Issuing the Order

In order to replenish the inventory and face the future demands of the final customer, the retailer must issue an order O_t to the supplier. The main problem faced by the retailer is that she does not know the future customer demand. Furthermore, the units ordered to the supplier in time period t would not arrive at the retailer until L time periods later.

Among the different policies that she may use for inventory management, we have discussed in Sect. 2.4 the Order-Up-To (OUT) policy. The goal of this policy is to face the customer demand with an inventory level s, also known as *base stock*. In Sect. 2.4, we have shown that s^* the optimal value of this base stock level has a closed expression if the demand on each period follows a normal distribution. In such case, it was shown that $s^* = \mu^L + z \cdot \sigma^L$, where μ^L is the expected demand over the L time periods, σ^L is the standard deviation of the customer demand over the L time periods and z is a factor (safety factor, as we know from Sect. 2.4) which usually depends on the ratio among the inventory costs and the stockout costs.

Since we have not made the assumption that the customer demand follows a normal distribution, there is no guarantee that the OUT policy is optimal. Nevertheless, it is a usual replenishment policy, and therefore, we will assume that it is adopted by the retailer. However, we will denote as s the so-obtained base stock in order to indicate that it does not have to be optimal. Furthermore, in general, the expected value and the standard deviation of the customer demand over L periods is not known, so it must be estimated by the supplier, possibly using demand historical data. If we denote by $\hat{d}^L$ and $\hat{\sigma}^L$ such estimates (we will discuss later how these estimates can be obtained), the base stock has the following expression:

$$s = \hat{d}^L + z \cdot \hat{\sigma}^L \tag{4.10}$$

Furthermore, it is to note that, in general, the so-computed base stock might be different for each time period t; therefore, its generic expression would be

$$s_t = \hat{d}_t^L + z \cdot \hat{\sigma}_t^L \tag{4.11}$$

where $\hat{d}_t^L$ and $\hat{\sigma}_t^L$ are know as the estimates of the expected final customer demand and its standard deviation over L periods, respectively, computed in time period t. Note also that, unless it is assumed that the inventory and/or stockout costs vary across the time periods, z is constant.

Once S_t is computed using one of the procedures discussed later in the book, the order to be issued in time period t can be computed using the following equation:

$$O_t = \max(s_t - I_t - W_t; 0) \tag{4.12}$$

Equation (4.13) reflects the fact that, in order to return the inventory level to s_t, we must order s_t units minus the units already in inventory (I_t) and minus the units that have been ordered but have not arrived yet (W_t). The equation also states that, in principle, it is not possible for the retailer to order negative quantities from the supplier.

Note that Eq. (4.13) introduces another non-linearity that, as in the previous case, can complicate the obtention of a simple model. Again, we can overcome this difficulty by assuming that the supplier can return the excess inventory without any cost, thus making it possible that O_t is negative. The implications of this assumption (returned orders) would be discussed later but, for the moment, it is clear that it simplifies the expression of the order, which is now:

$$O_t = s_t - I_t - W_t \tag{4.13}$$

Looking at the difference between two consecutive time periods t and $t-1$, the following equation holds:

$$O_t - O_{t-1} = s_t - s_{t-1} - (I_t - I_{t-1}) - (W_t - W_{t-1}) \tag{4.14}$$

Taking into account Eq. (4.7) and Eq. (4.9) to substitute in Eq. (4.14) the difference of inventories and the difference of work in process, we obtain the following expression:

$$O_t = s_t - s_{t-1} + D_t \tag{4.15}$$

Equation (4.15) is the key to obtain a closed expression of the BWE indicator, since s_t is obtained from demand historical data, we have that, effectively, the order in time period t can be written as a function of the current and past customer demand, i.e. $O_t = f(D_t, D_{t-1}, \ldots)$. Furthermore, the $NSAmp$ indicator requires expressing the inventory in terms of the customer demand. Therefore, taking into account Eq. (4.4) that expresses the work in process as a function of the orders and substituting in Eq. (4.13), we have

$$I_t = s_t - \sum_{i=1}^{L-1} O_{t-i} - O_t = s_t - \sum_{i=0}^{L-1} O_{t-i} \tag{4.16}$$

Therefore, it is clear that once O_t is written as a function of the final customer demand, it is also possible to write I_t in terms of the final customer demand.

Let us recall the assumptions that have led us to obtain this expression, as some of them would be relaxed later:

- The supplier uses a base stock policy for inventory replenishment.
- The supplier can return to the provider the excess inventory without additional costs (returned orders).

To determine the function linking the orders with the demand, we have to decide how specifically the supplier estimates the average and standard deviation of the final customer demand as well as the behaviour of the final customer demand. These aspects are addressed in Sect. 4.5.4.

4.5.4 *Demand Estimates*

To develop the models in this section, two different hypotheses will be done regarding how the retailer conducts the forecast of the final customer demand. In the first case—discussed in Sect. 4.5.4.1—we assume the best possible situation, i.e. the retailer is able to produce a forecast with the mean minimum error (measured as MSE, see Sect. 2.3.4) on the long run. In the second case—discussed in Sect. 4.5.4.2—we assume that the retailer produces a forecast using some standard forecasting method (in this case, the moving average). This distinction will provide some interesting insights regarding the value of the bullwhip effect and how it is affected by the method adopted to forecast the final customer demand.

4.5.4.1 MMSE Estimation

In this case, two sub-cases can be distinguished. The first sub-case considers that there is no correlation of the demand across periods ($\rho = 0$) whereas in the second sub-case, we deal with the most general case.

If the demand is i.i.d. and it is assumed that a MMSE method is employed to estimate it, then it turns out that the estimate of the demand on each period is μ the expected value of the demand. As a consequence, the estimate of the demand across L periods does not depend on t either. Furthermore, the error in the estimate is constant and does not depend on each period; therefore, $s_t = s_{t-1}$, see the expression in Eq. (4.11) and Eq. (4.15) is simplified to

$$O_t = D_t \tag{4.17}$$

In other words, if the demands are independent across periods and an optimal estimation method is employed, using the OUT policy turns out to be simply ordering in each time period t the demand in this period. This policy is known as *chase* policy, as it simply tries to restore the inventory levels to the value it had before satisfying the final customer demand.

If the chase policy is adopted, then it is clear that the BWE indicator is 1 (i.e. there is no amplification of the variance of the orders with respect to that of the demand). Regarding the indicator $NSAmp$, according to Eq. (4.16), it holds the following:

$$NSAmp = \frac{V[I_t]}{D[D_t]} = L \tag{4.18}$$

What the equations express is that, intuitively, the chase policy does not create any variance amplification in the orders—after all, it mimics the behaviour of the demand so it is not possible for the orders to have a different variance than that of the demand—but it does create an amplification in the level of inventory, due to the fact that the lead time is not zero and the inventory has to be protected during more time periods.

Let us move to the case where the final customer demands are not independent, but present some correlation across time periods. More specifically, we will assume that the demand can be modelled as the well-known AR(1) time series. Then, we know that the MMSE estimate in time period t of the demand occurring in time period $t+h$ is the following, see Eq. (2.24) in Sect. 2.3.4:

$$\hat{d}_t(h) = \frac{d}{1-\rho} + \rho^h\left(D_t - \frac{d}{1-\rho}\right) \tag{4.19}$$

As a consequence, $\hat{d}_t^L$ the estimate across L periods is

$$\hat{d}^L(t) = \sum_{h=1}^{L} \hat{d}_t(h) = \sum_{h=1}^{L} \frac{d}{1-\rho} + \rho^h\left(D_t - \frac{d}{1-\rho}\right) = \tag{4.20}$$

$$L\frac{d}{1-\rho} + \rho\frac{1-\rho^L}{1-\rho}\left(D_t - \frac{d}{1-\rho}\right) \tag{4.21}$$

Therefore, the expression in Eq. (4.15) is

$$O_t = D_t + \rho\frac{1-\rho^L}{1-\rho}(D_t - D_{t-1}) + z(\hat{\sigma}_t^L - \hat{\sigma}_{t-1}^L) \tag{4.22}$$

In order to simplify the expression above, let us denote the geometric sum of ρ across L as K, i.e.

$$K = \sum_{i=1}^{L} \rho^i = \rho\frac{1-\rho^L}{1-\rho} \tag{4.23}$$

Furthermore, since we are assuming a AR(1) model, we know that $V[e_t^L]$ does not depend on t, see Eq. (2.35) in Chap. 2. Therefore, $\hat{\sigma}_t^L = \hat{\sigma}_{t-1}^L$ and, as a consequence, Eq. (4.22) can be expressed as

$$O_t = (1+K)\cdot D_t - K\cdot D_{t-1} \tag{4.24}$$

Taking variances in the above expression,

$$V[O_t] = (1+K)^2\, V[D_t] + K^2\, V[D_{t-1}] - 2\,K(1+K)cov[D_t, D_{t-1}]$$

and recalling that $cov[D_t, D_{t-1}] = \rho V[D]$, we have

$$V[O_t] = V[D]\left((1+K)^2 + K^2 - 2\rho K(1+K)\right) \tag{4.25}$$

Therefore, the BWE is

$$BWE = 1 + 2K(1+K)(1-\rho) \tag{4.26}$$

or, substituting back K,

$$BWE = 1 + 2\rho \cdot (1 - \rho^L)\frac{1 - \rho^{L+1}}{1 - \rho} \tag{4.27}$$

If K is substituted directly in Eq. (4.25), the equivalent expression is

$$BWE = \frac{1}{(1-\rho)^2} \cdot [(1 - \rho^{L+1})^2 + \rho^2(1 - \rho^L)^2 - 2\rho^2(1 - \rho^{L+1})(1 - \rho^L)] \tag{4.28}$$

Note that Eq. (4.28) is quite cumbersome and it will be used only in the context of comparing the BWE indicator in different echelons in the SC carried out in Sect. 4.6.1. In some texts, the expression of the BWE in Eq. (4.27) appears rearranged with a common denominator, i.e.

$$BWE = \frac{(1 + \rho)(1 - 2\rho^{L+1}) + 2\rho^{2L+2}}{1 - \rho} \tag{4.29}$$

Finally, another alternative formulation allows for a more compact expression of BWE: If we take Eq. (4.27) and regroup the terms, we have

$$BWE = 1 + 2 \cdot (1 - \rho^{L+1})\frac{\rho - \rho^{L+1}}{1 - \rho} \tag{4.30}$$

Taking into account that the last fraction is simply $\sum_{i=1}^{L} \rho^i$, we can write

$$BWE = 1 + \sum_{i=1}^{L} 2(1 - \rho^{L+1})\rho^i \tag{4.31}$$

From these equivalent expressions, we can obtain some insights regarding the bullwhip effect:

- The bullwhip effect (demand amplification) exists if and only under positively correlated demand ($\rho > 0$). This can be proved using the second expression of BWE in Eq. (4.29):

$$BWE \leq 1 \Leftrightarrow \frac{(1 + \rho)(1 - 2\rho^{L+1}) + 2\rho^{2L+2}}{1 - \rho} \tag{4.32}$$

$$\Leftrightarrow 1 - 2\rho^{L+1} + \rho - 2\rho^{L+2} + 2\rho^{2L+2} \leq 1 \Leftrightarrow \tag{4.33}$$

$$2\rho\left((1 - \rho^l) - (\rho^{L+1} - \rho^{2L+1})\right) \leq 0 \Leftrightarrow \tag{4.34}$$

$$2\rho(1 - \rho^l)(1 - \rho^{L+1}) \leq 0 \Leftrightarrow \rho \leq 0 \tag{4.35}$$

This is an interesting result since, as we already discussed in Chap. 2, there are many empirical evidence suggesting that many final products present a positively correlated demand.

- For a positively correlated demand, BWE is monotonically increasing with L: the longer the lead times, the higher the bullwhip effect. This can be easily verified by looking the expression of BWE in Eq. (4.31).
- For a positively correlated demand, BWE is non-monotonically increasing with ρ. Indeed, it can be shown that there is a value ρ^* for which the BWE is maximum. Perhaps, the simplest way to show it is to express Eq. (4.31) as the sum of the following function $f_i(\rho, L) = 2(1 - \rho^{L+1})\rho^i$. By doing so, it is clear that

$$BWE = 1 + \sum_{i=1}^{L} f_i(\rho, L) \tag{4.36}$$

We can analyse the behaviour of the functions $f_i(\rho, L)$ with respect to ρ. It is clear that

$$\frac{\partial f_i(\rho, L)}{\partial \rho} = 2 \cdot i \cdot \rho^{i-1}(1 - \rho^{L+1}) - 2 \cdot \rho^i \cdot (L+1) \cdot \rho^L = \tag{4.37}$$

$$2\rho^{i-1}\left[i - (i + L + 1)\rho^{L+1}\right] \tag{4.38}$$

By making the above expression equal to zero, it can be seen that the functions $f_i(\rho, L)$ have the following maximum in ρ_i^*:

$$\rho_i^* = \left(\frac{i}{i + L + 1}\right)^{\frac{1}{L+1}} \tag{4.39}$$

It can be seen that it is a maximum since $\frac{\partial f_i(\rho,L)}{\partial \rho} > 0$ in the interval $(0, \rho_i^*)$ (recall that we assume that the demand is positively correlated) while in the interval $(\rho_i^*, 1)$, we have that $\frac{\partial f_i(\rho,L)}{\partial \rho} < 0$. Furthermore, note that ρ_i^* is an increasing function with respect to i.

From these results, it is easy to show that BWE is an increasing function in the interval $(0, \rho_1^*)$, with $\rho_1^* = \left(\frac{1}{L+2}\right)^{\frac{1}{L+1}}$: this is confirmed since all $f_i(\rho, L)$ are increasing functions in the interval $(0, \rho_i^*)$ and in view that ρ_i^* is increasing with i. Furthermore, it is possible to show that BWE is concave in the interval $(\rho_1^*, 1)$. Figure 4.2 shows the different values of BWE as a function of ρ (for its non-negative domain) and for several values of L. As it can be seen, the impact of the lead times is small for low levels of correlation, whereas it is high for high levels of correlation (Fig. 4.3).

The same logic can be applied to obtain $NSAmp$. However, in this case, the expression obtained is quite ugly and few insights can be derived. Regarding BWE, however, the expression in Eq. (4.27) indicates that, if $\rho > 0$ (the demand presents

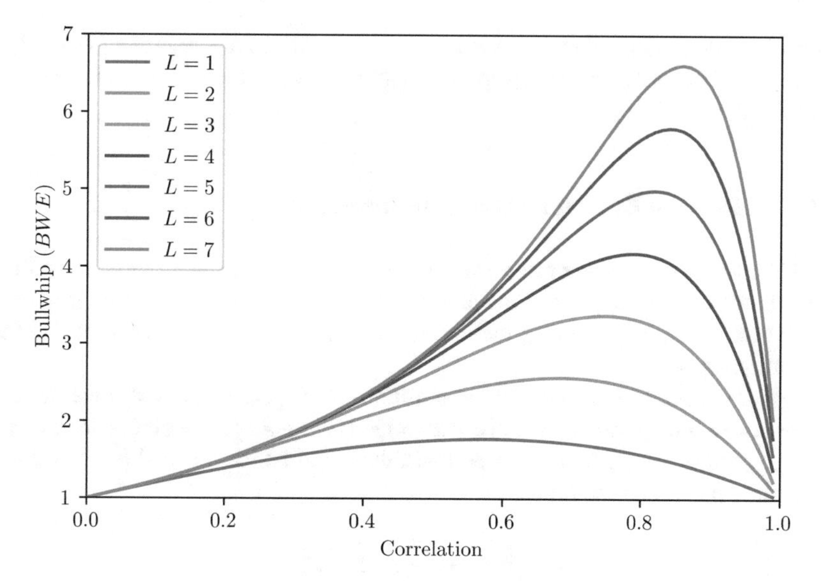

Fig. 4.2 Values of BWE for a positively correlated demand

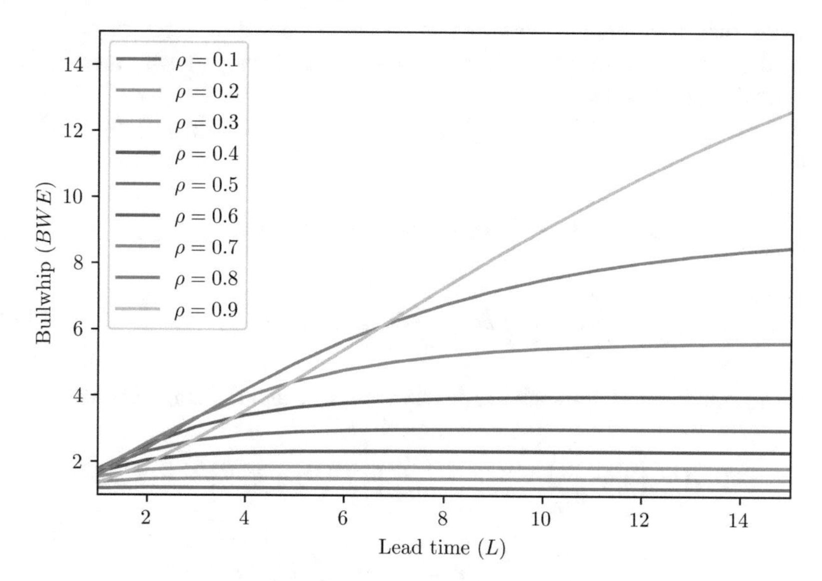

Fig. 4.3 Values of BWE for a positively correlated demand

a positive correlation), then there is an amplification of the variance of the orders. However, whenever such correlation is negative ($\rho < 0$), the variance of the orders is reduced.

4.5.4.2 Demand Estimation Using the Moving Average

Next, we will study the case where the demand is estimated using the moving average. Apart from being a simple method employed in real-life settings, we will use it as a proxy of a typical approximate method for demand forecast (instead of the MMSE estimation assumed before).

Again, we need to estimate $\hat{d}_t^L$; therefore, in time period t, the average demand of the last m time periods would be computed (being m a parameter of the moving average), and it would be forecasted that this would be the demand in each one of the L time periods in the future, i.e.

$$\hat{d}_t^L = L \cdot \frac{\sum_{i=0}^{m-1} D_{t-i}}{m}$$

(Note, that, according to the event sequence discussed in Sect. 4.5, at the time where the forecast has to be done, the demand D_t is already known; therefore, the summation starts in zero).

Therefore, substituting in Eq. (4.15) the values of s_t according to Eq. (4.11), we have

$$O_t = D_t + \frac{L}{m}(D_t - D_{t-m}) + z(\hat{\sigma}_t^L - \hat{\sigma}_{t-1}^L) =$$

$$= \left(1 + \frac{L}{m}\right) D_t - \frac{L}{m} D_{t-m} + z(\hat{\sigma}_t^L - \hat{\sigma}_{t-1}^L) \tag{4.40}$$

Taking variances to compute the demand variance amplification, we have

$$V[O_t] = \left(1 + \frac{L}{m}\right)^2 V[D] + \left(\frac{L}{m}\right)^2 V[D] + z^2\, V[\hat{\sigma}_t^L - \hat{\sigma}_{t-1}^L] +$$

$$2\left(1 + \frac{L}{m}\right)\left(\frac{L}{m}\right) cov[D_t, D_{t-m}] +$$

$$+2\left(1 + \frac{L}{m}\right) z \cdot cov[D_t, \hat{\sigma}_t^L - \hat{\sigma}_{t-1}^L] + 2\frac{L}{m} z \cdot cov[D_{t-m}, \hat{\sigma}_t^L - \hat{\sigma}_{t-1}^L] \tag{4.41}$$

Since it is assumed that the demand is AR(1), we know that $cov[D_t, D_{t-m}] = \rho^m V[D]$. Furthermore,[1] it can be shown that, in these conditions, $cov(D_{t-i}, \hat{\sigma}_t^L]) =$

[1] The proof is discussed in Sect. 4.9.

$0 \ \forall i = 1, \ldots, m$. As a consequence, the terms in the last line of Eq. (4.41) are zero, and the final expression is

$$V[O_t] = \left[\left(1 + \frac{L}{m}\right)^2 + \left(\frac{L}{m}\right)^2 + 2\left(1 + \frac{L}{m}\right) \cdot \left(\frac{L}{m}\right) \rho^m \right] V[D] + z^2 \, V[\hat{\sigma}_t^L - \hat{\sigma}_{t-1}^L]$$

or

$$V[O_t] = \left(1 + \left(2\frac{L^2}{m^2} + 2\frac{L}{m}\right)(1 - \rho^m)\right) V[D] + z^2 \, V[\hat{\sigma}_t^L - \hat{\sigma}_{t-1}^L]$$

Since the variance is always positive, it is clear that

$$BWE = \frac{V[O_t]}{V[D]} \geq 1 + \left(2\frac{L^2}{m^2} + 2\frac{L}{m}\right)(1 - \rho^m) \quad (4.42)$$

The following conclusions can be derived from Eq. (4.42):

- BWE increases with L, i.e. the order variability amplification is higher as the lead times increase. The ultimate reason is that, as the lead time increases, there is a need to protect the inventories from the demand uncertainties for a longer time period. Such uncertainty is higher as the lead time to estimate the demand is higher.
- As m increases, the order variability decreases. The rationale is that m serves to smooth the demand and, since the demand is stationary, the longer the number of points, the more stable the forecast.
- There is demand amplification even for the case of uncorrelated demand ($\rho = 0$ or iid customer demand), in contrast with the case of the MMSE forecast. This can be seen as the order variability amplification caused by using a non-optimal forecast method.
- If $\rho > 0$ (positively correlated demand), then the variability of the amplification decreases. The explanation is that, if there is a strong positive correlation, then it is easier to make better forecasts.
- If $\rho < 0$ (negatively correlated demand), then BWE increases or decreases depending on whether m is even or not. The explanation of this surprising result is given by the fact that, in the case of negatively correlated demand, time periods with (probabilistically) high demand are followed by time periods with (probabilistically) low demand. Therefore, an odd value of m means that, when estimating the future demand, there is always one more time period with high (low) demand than the number of time periods with low (high) demand. Therefore, such a forecast is not good and the bullwhip effect increases. In contrast, if m is even, such alternated high and low values are adequately captured and the bullwhip effect is reduced.

4.5.5 A Simple Simulation Model

The sets of equations developed in the previous sections can be also easily plugged into a spreadsheet or coded in a programming language, thus having a simulation model of the bullwhip effect. We will not use the model in this section, but it will be useful for the next chapters, particularly in Chap. 6 where we will remove some of the hypotheses of our original model and this will introduce non-linearities that will make it infeasible to develop close-formula models. Just as a reference, we give below the code of such a program in Python. Note that we show only the sketch of the main function and a function to estimate the demand (in this case using the moving average) and do not give further code lines to print or analyse the results.

Basic node routine

```
import numpy as np
import random

# estimates demand for next period based on a m-periods moving average
def moving_average(D, m, t):
  last_sample = t
  # at the beginning of the simulation there are no sufficient data
  # for the demand:
  first_sample = max(t-m,0)+1
  sum = 0
  for x in range(first_sample, last_sample+1):
  sum = sum + D[x]
  estimate = sum/(last_sample - first_sample+1)
  return estimate


# main function:
# time parameters
t_max = 500 # number of simulations

#supply chain decisions
L = 5 # lead time
# demand initialisation and parameters
D = [0]
mu = 100
sigma = 5

# state variables of the node
R = [0]
O = [0]
I = [100]
W = [0]

# decisional parameters:
# parameters of the order-up-to
```

```
s = 10
t = 1
#parameters of the moving average
m = 3

# starting the simulation
while t <= t_max:
  # the retailer receives the order placed L periods before
  if(t - L >= 0):
    R.append(O[t-L])
  else:
    R.append(0)
  # the inventory is increased
  I.append(I[t-1] + R[t])
  # work in process is decreased
  W.append(W[t-1] - R[t])
  # customer demand is generated
  curr_demand = round(random.gauss(mu, sigma),0)
  D.append(curr_demand)
  # customer demand is statisfied
  I[t] = I[t] - curr_demand

  #calculates current order (moving average of m periods)
  curr_order = moving_average(D,m,t)*L - I[t] - W[t]
  O.append(curr_order)

  # update work in process (increases with the current order)
  W[t] = W[t] + curr_order
  # increases current period
  t = t + 1
```

Note that, in contrast to the analytical models developed in the previous section, the correct choice of the parameters (particularly the length of the simulation `t_max` has a great influence on the results. It is then convenient to use a large value of `t_max` as well as to discard the values obtained in the first periods to produce long-run values that are consistent with those obtained in the analytical formulae.

4.6 The Model at Supply Chain Level

The model that we have discussed so far only takes into account the interaction between the supplier and the retailer. However, the model can be extended to represent a SC with several echelons and represent the variability of the demand perceived by each node across the SC. More specifically, we can use σ_i to denote the standard deviation experienced by the demand received by the customer of echelon i in the supply chain. Since, for echelon i, $\sigma_i = \sqrt{V[D_i]}$ and $\sqrt{V[O_i]}$ is the standard deviation of its demand as perceived by node $i + 1$, we can write

$$\sigma_{i+1} = \sigma_i \cdot \sqrt{BWE_i} \tag{4.43}$$

where BWE_i is the value of BWE for node i. In this manner, we can write recursively the expression of the standard deviation of the demand observed by a node i in the SC, i.e.

$$\sigma_{i+1} = \Pi_{k=1}^{i} \sigma \cdot \sqrt{BWE_k} \tag{4.44}$$

where σ is the standard deviation of the demand of the final customer of the SC. By using this expression, we can study the propagation of the bullwhip effect in the SC also depending on the hypotheses adopted to obtain BWE_k.

We start with the expression of BWE obtained with the hypothesis of MMSE estimation. For simplicity, we assume that the lead time in all nodes is the same (L), so Eq. (4.44) is

$$\sigma_{i+1} = \sigma^i \cdot \left(1 + 2\rho(1-\rho^L)\frac{1-\rho^{L+1}}{1-\rho}\right)^{i/2} \tag{4.45}$$

and we can depict the values of σ_i for different cases. For instance, in Fig. 4.4, we give the values of σ_i for the case of $\rho = 0.5$, different values of the lead time and assuming that the standard deviation of the demand of the final customer is 1. As it can be seen, if the lead time is not low (e.g. $L = 4$), the amplification of the standard deviation is notable for the last nodes in the SC (in this case, it is around 10 for $i = 4$ and more than 20 for $i = 6$). This is 10 and 20 times greater than the standard deviation of the final customer demand, which clearly is inducing extra costs in the nodes upstream in the SC. Note that four (or six) nodes are not unrealistic in many situations, as in many SCs, there are manufacturers, wholesalers, distributors, resellers, retailers, etc.

If we assume that all nodes perform a forecast based in the moving average using m periods, the expression in Eq. (4.44) can be written as

$$\sigma_{i+1} \geq \sigma^i \cdot \left[1 + \left(2\frac{L^2}{m^2} + 2\frac{L}{m}\right)(1-\rho^m)\right]^{i/2} \tag{4.46}$$

In this case, Eq. (4.46) can give only a lower bound of the standard deviation, which is depicted in Fig. 4.5 for $\rho = 0.5$, $\sigma = 1$ and $m = 5$ for different values of the lead time. As it can be seen, the curve is much more steep than in Fig. 4.4, which makes sense if we consider that the error of the MA forecast would be higher (Fig. 4.6).

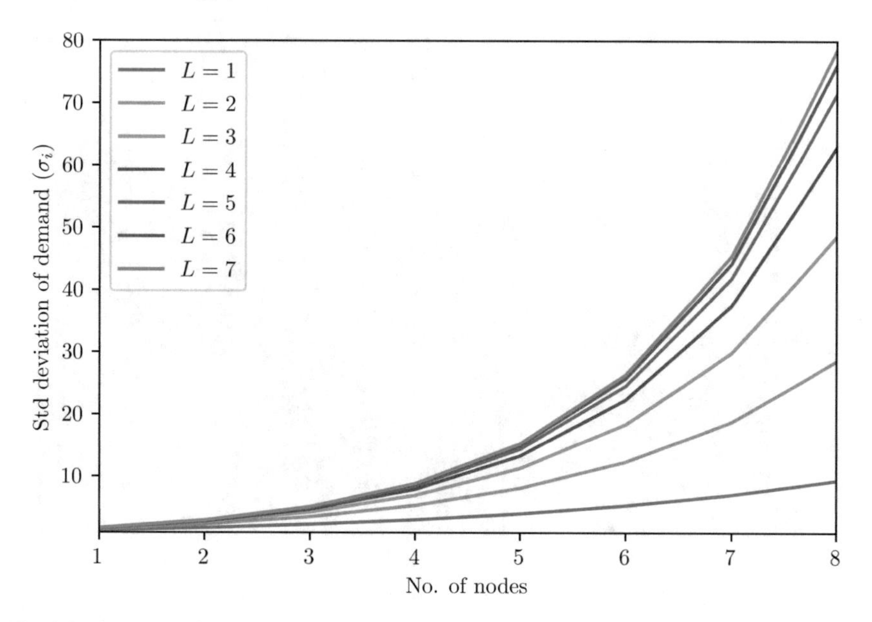

Fig. 4.4 Evolution of the standard deviation of the demand perceived by a node in a SC (MMSE estimation, $\rho = 0.5$ and $\sigma = 1$), depending on the lead time L between the nodes

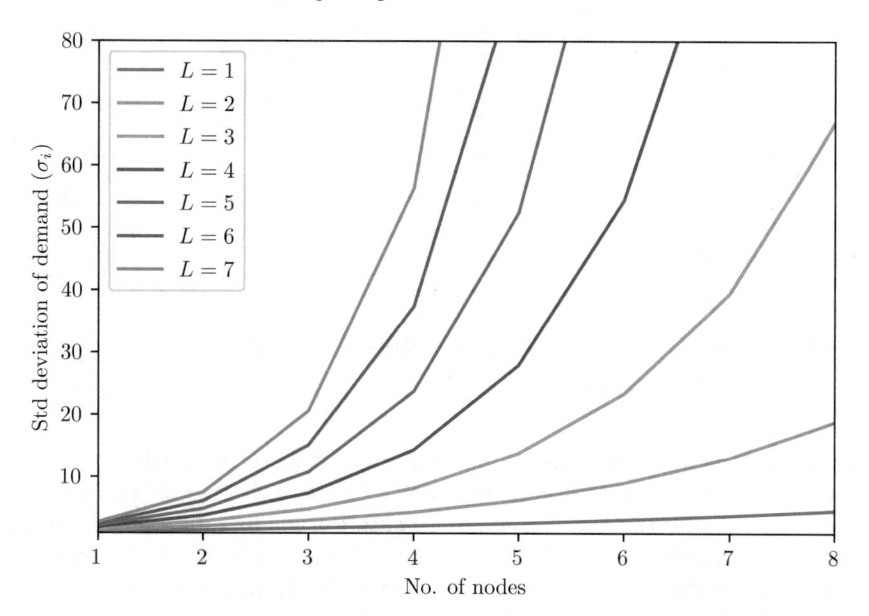

Fig. 4.5 Evolution of the standard deviation of the demand perceived by a node in a SC (MA estimation with $m = 5$, $\rho = 0.5$ and $\sigma = 1$), depending on the lead time L between the nodes

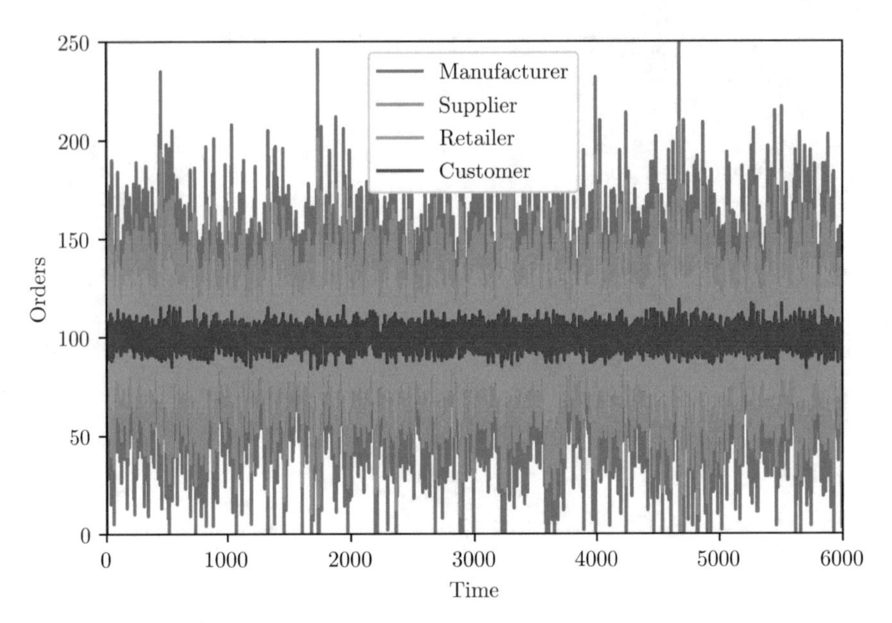

Fig. 4.6 Simulation of the variability amplification at SC level: basic model

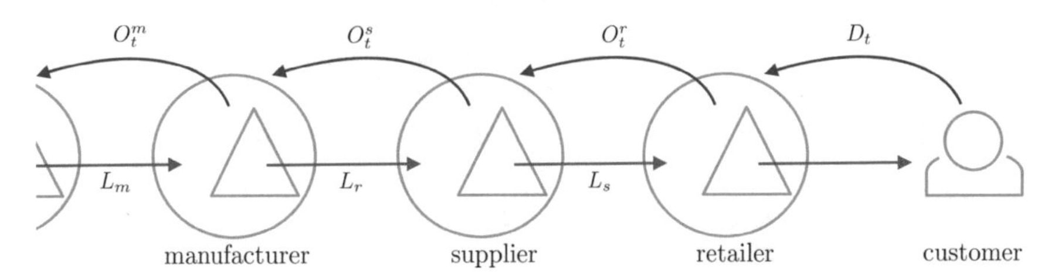

Fig. 4.7 Three-echelon SC

4.6.1 An Expression of the BWE for a Two- and Three-Echelon SC

Even if it somewhat involving and the equations obtained are cumbersome, it is possible to extend the expressions of the BWE indicator observed by each node of the SC depending on the demand of the final customer and for different lead times if an MMSE forecast is assumed. In this section, we will provide the expressions and give the indications to derive them, even if we do not provide a step-by-step proof and we will give the references to the reader in Sect. 4.9.

More specifically, we will give the expressions for BWE in each node for the three-echelon SC depicted in Fig. 4.7 composed of a retailer, a supplier and a manufacturer. Each node performs an MMSE estimation of the demand observed by the node and adopts an OUT policy to replenish the inventory. The lead times among each node are different, and it is denoted by L_r, L_s and L_m for the retailer, supplier and manufacturer, respectively. As in the previous models, we assume that the final customer demand can be modelled using an AR(1) time series and that the returns can be negative.

Under the assumptions of MMSE estimation and OUT policy, we know that the expression of the orders of a node as a function of the demand observed by a node is given by Eq. (4.24):

$$O_t = (1 + K) \cdot D_t - K \cdot D_{t-1} \tag{4.47}$$

Taking into account that in Eq. (4.24), $K = \rho \frac{1-\rho^L}{1-\rho}$, the expression can be written as

$$O_t = \frac{1-\rho^{L+1}}{1-\rho} \cdot D_t - \frac{1-\rho^L}{1-\rho} \cdot D_{t-1} = \frac{1}{1-\rho}\left[(1-\rho^{L+1})D_t - (1-\rho^L)D_{t-1}\right] \tag{4.48}$$

The previous equation can be formulated for each node in the SC taking into account that O_t are the orders issued by the node and D_t the demand seen by the node. Therefore, for the retailer, we can write

$$O_t^r = \frac{1}{1-\rho}\left[(1-\rho^{L_r+1})D_t - (1-\rho^{L_r})D_{t-1}\right] \tag{4.49}$$

while for the supplier, the expression of the orders is

$$O_t^s = \frac{1}{1-\rho}\left[(1-\rho^{L_s+1})O_t^r - (1-\rho^{L_s})O_{t-1}^r\right] \tag{4.50}$$

Finally, for the manufacturer, it is

$$O_t^m = \frac{1}{1-\rho}\left[(1-\rho^{L_m+1})O_t^s - (1-\rho^{L_m})O_{t-1}^s\right] \tag{4.51}$$

As it can be seen, it is possible to substitute O_t^r and O_{t-1}^r in Eq. (4.50) by its expression in Eq. (4.49) so we obtain

$$O_t^s = \frac{1}{(1-\rho)^2} \cdot [(1-\rho^{L_s+1})(1-\rho^{L_r+1})D_t - \tag{4.52}$$

$$\left((1-\rho^{L_{s+1}})(1-\rho^{L_r}) + (1-\rho^{L_s})(1-\rho^{L_{r+1}})\right) D_{t-1} + \tag{4.53}$$

$$(1-\rho^{L_s})(1-\rho^{L_r})D_{t-2}] \tag{4.54}$$

In this expression, it is possible to take variances and obtain $V[O_t^s]$, taking into account also the correlation among the demand. It is quite involving as the expressions are quite long and require further simplification. The final expression is

$$V[O_t^s] = \frac{\sigma^2}{(1-\rho)^2 \cdot (1-\rho^2)} \cdot [(1-\rho^{L_s+L_r+1})^2 + \rho^2(1-\rho^{L_s+L_r})^2 - \tag{4.55}$$

$$-2\rho^2(1-\rho^{L_s+L_r+1})(1-\rho^{L_s+L_r})] \tag{4.56}$$

and the expression for the bullwhip indicator for the supplier (denoted as BWE_s) is

$$BWE_s = \frac{1}{(1-\rho)^2} \cdot [(1-\rho^{L_s+L_r+1})^2 + \rho^2(1-\rho^{L_s+L_r})^2 - \tag{4.57}$$

$$-2\rho^2(1-\rho^{L_s+L_r+1})(1-\rho^{L_s+L_r})] \tag{4.58}$$

A similar procedure can be adopted to obtain BWE_m the bullwhip indicator for the manufacturer, which is

$$BWE_m = \frac{1}{(1-\rho)^2} \cdot [(1-\rho^{L_s+L_r+L_m+1})^2 + \rho^2(1-\rho^{L_s+L_r+L_m})^2 - \tag{4.59}$$

$$-2\rho^2(1-\rho^{L_s+L_r+L_m+1})(1-\rho^{L_s+L_r+L_m})] \tag{4.60}$$

Taking into account also the expression for the bullwhip effect in the retailer from Eq. (4.28), we also have that

$$BWE_r = \frac{1}{(1-\rho)^2} \cdot [(1-\rho^{L+1})^2 + \rho^2(1-\rho^{L})^2 - 2\rho^2(1-\rho^{L+1})(1-\rho^{L})] \tag{4.61}$$

As we can see from Eqs. (4.61), (4.57) and (4.59), we see that the generic expression for the bullwhip effect observed by an echelon i is

$$BWE_i = \frac{1}{(1-\rho)^2} \cdot [(1-\rho^{L(i)+1})^2 + \rho^2(1-\rho^{L(i)})^2 - 2\rho^2(1-\rho^{L(i)+1})(1-\rho^{L(i)})] \tag{4.62}$$

where $L_(i)$ indicated the accumulated lead time across all the nodes downstream node i plus the lead time of the supplier of node i.

Equation (4.62) explains the amplification of the variability across the SC: it is $L(i)$ the sum of the accumulation of all downstream replenishment lead times plus the local replenishment lead time the factor influencing the variance of order rates. $L(i)$ is strictly increasing with i. Regarding BWE_i, if $\rho > 1$, then BWE_i is also increasing with $L(i)$. It can be also seen that how the lead times are distributed among the nodes is not relevant for the node.

4.7 Implications of the Model: The Causes of Bullwhip Effect Revisited

The last model seen in the previous section serves to explain, even if at a basic level, some potential areas for reducing/removing the propagation of the bullwhip effect across the supply chain, in line with some of the actions discussed in Sect. 3.5 in Chap. 3:

1. **Lead-time reduction**. The effect of lead time appears as the main factor influencing the bullwhip effect. Only in the case of uncorrelated demand *and* an MMSE demand estimation, the lead time does not play any role in increasing the bullwhip effect. Since the lead time is related to the time that occurred between ordering and replenishment, it is clear that this delay influences the SC dynamics unless both the demand and the forecasting are 'time-independent'.
   In general, lead time is constituted by several activities:

   - Time to order. This is the time from the instant where the retailer detects the need to replenish the inventory until the order is actually launched to the supplier. This time is related to the time between replenishments or cycle time. In most companies, the orders are not launched immediately due to the fixed ordering costs. As it is well-known from basic inventory management policies, the higher the fixed costs, the bigger the order size and the longest the time between orders. Therefore, one way to reduce the time to order is to reduce the fixed ordering costs.
   - Time for the provider to manufacture the product. In some cases, it might be assumed that the provider operates under a MTS (Make-To-Stock) policy so this time is zero; however, in some cases, some pre-assembly or even the full manufacture of the product has to be accomplished.
   - Time to prepare the order and to ship it to the customer.

2. **Information transparency**, also known as *demand integration*. As we can see in Figs. 4.4 and 4.5, which relay on the SC-wide expressions for the bullwhip effect discussed in Sect. 4.6, the number of nodes increases the bullwhip effect, so one idea is that the demand forecast for node i is done based on the demand of the final customer rather than in the demand perceived by the node (i.e. the orders issued by node $i-1$). By doing so, the effect—in terms of amplification of bullwhip across the SC—is equivalent to reduce the number of echelons in the SC. Clearly, one way is that the retailer shares the POS data with the rest of the nodes of the SC in line with the approaches to mitigate the bullwhip effect discussed in Sect. 3.5. An alternative to remove the need that each node in the SC performs a demand forecast is initiatives such as VMI or CFPR also discussed in Sect. 3.5. Finally, from Eqs. (4.45) and (4.46), the effect of requiring multiple nodes in the SC to deliver the product to the final customer can be easily appreciated. Strategies such as e-Shopping, in which the distribution network is bypassed and the information and materials flow directly between the end consumer and the product suppliers, also can play a role in reducing the bullwhip effect.

Upon the question of which of these two aspects (lead-time reduction or information transparency) should be given the highest priority, there is no clear answer, even if it has been traditionally discussed that there are limits in the information transparency (as we have seen in Sect. 4.5, the amplification of the variability of the orders and the inventory also occurs among two nodes of the SC).

4.8 Summary

In this chapter, a basic, two-echelon model to understand the SC dynamics has been developed step by step. The different hypotheses employed in developing the model have been discussed, as well as the implications in the results. The model has been extended to the SC, and it has served us to revisit the causes of the bullwhip effect discussed in Chap. 3, along with its remedies. In the next chapters, the basic model discussed here will be extended to contemplate additional effects or to remove some of the hypotheses that have been formulated.

4.9 Further Readings

The assumption that the excess inventory can be returned made in 4.5.3 is a common hypothesis in the SC literature using analytical models, see, e.g. [1–3].

The derivation of the MMSE expression for the bullwhip effect has been taken from [4], along with the formulae of BWE as a geometric sum. The expression of the bullwhip effect when the forecast is the moving average is in [1], although in Sect. 4.5.4.2, we have mainly followed the reasoning in [5] to derive the corresponding expressions. In this section, we have assumed that the covariance of the demand and the standard deviation of the error estimated are zero, a lemma that can be found in [5] with the proof given in [6] or in [7]. A similar, detailed derivation of the expression can be found in [8]. While in the chapter, it has been assumed that the demand can be modelled as a AR(1) series, in [9] it is shown that the bullwhip effect appears also in a general ARIMA model of the demand. Finally, in [10, 11], the *anti-bullwhip* effect (i.e. the reduction of the order variability across the SC) is discussed, and the demand models leading to this phenomenon are identified.

The effect of the lead time in worsening SC dynamics is known since the first studies by Forrester [12] and, as early as 1991, [13] show in that year that shortening lead times causes a notable reduction in the peak amplification. In [14], however, it is shown that increasing lead times do not necessarily have to increase the bullwhip effect and that there are demand models for which this fact does not hold. In [15], it is shown that the cyclical oscillation of on-hand and on-order inventory does not require that the demand is random, thus pointing at the exogenous causes of demand variability amplification. The fact that the bullwhip effect cannot be removed by centralised forecasting and ordering as long as the demand is correlated and the lead times are not zero is called by [3] the 'core bullwhip effect', while other factors affecting the bullwhip effect—such as batching or inaccurate order placements—are denoted as 'incremental bullwhip'.

Aside to the expressions of BWE derived in Sect. 4.5.4 for the MMSE and MA estimates, there are additional expressions derived for different forecasting techniques, such as the exponential smoothing, or the demand has a linear trend, see [16], or in the case that seasonality exists (see, e.g. [17–20]). A relatively recent reference addressing the differences of forecasting techniques on the bullwhip effect is [21], while the effect of temporal aggregation to produce less bullwhip-prone forecast decisions is investigated in [22]. Similarly, the effect of different data aggregation is studied in [23].

The derivation of ordering policies aimed at reducing the bullwhip effect has also received a lot of attention, see, e.g. [24, 25]. Furthermore, simulation studies have been conducted to experiment with more sophisticated forecasting and ordering policies [26]. The amplification of the bullwhip effect across the SC is discussed in [5], where Eq. (4.46) is derived. The expressions of the BWE indicators for the three-echelon SC are obtained in [27], where equivalent expressions are also obtained for the $NSAmp$ indicator. In [28], a new perspective of this amplification is offered by modelling the SC using differential equations and finding the equilibrium states of the nodes. The upper bounds of this amplification across the nodes are addressed in [29]. Some of the aforementioned references use a *decomposition* approach in which the SC is decomposed into two-node pairs, an approach which is challenged in [30] as it is shown to underestimate the bullwhip effect in the whole SC. In [31], the effect of e-business strategies on the SC dynamics is modelled and discussed, whereas some empirical recollections on the impact of ICT in the SC can be found in [32–34].

Aside to the analytical models described here, early approaches to quantify the bullwhip effect have been developed using a control theory approach [35] or discrete-event simulation [36]. The combination of simulation and optimisation (most of the time using approximate methods) has been used in the literature to establish the best combination of operational parameters in the SC (typically the parameters of the forecasting techniques and or those of the replenishment policies). These contributions include [37–43], among others.

References

1. Lee, H., Padmanabhan, V., Whang, S.: Information distortion in a supply chain: the bullwhip effect. Manag. Sci. **43**(4), 546–558 (1997)
2. Lee, H., So, K., Tang, C.: Value of information sharing in a two-level supply chain. Manag. Sci. **46**(5), 626–643 (2000)
3. Sodhi, M., Tang, C.: The incremental bullwhip effect of operational deviations in an arborescent supply chain with requirements planning. Europ. J. Operat. Res. **215**(2), 374–382 (2011)
4. Luong, H.: Measure of bullwhip effect in supply chains with autoregressive demand process. Europ. J. Operat. Res. **180**(3), 1086–1097 (2007)

5. Chen, F., Drezner, Z., Ryan, J., Simchi-Levi, D.: Quantifying the bullwhip effect in a simple supply chain: the impact of forecasting, lead times, and information. Manag. Sci. **46**(3), 436–443 (2000)
6. Ryan, J.: Analysis of inventory models with limited demand information. Analysis of Inventory Models with Limited Demand Information (1997)
7. Chen, F., Drezner, Z., Ryan, J., Simchi-Levi, D.: Quantifying the bullwhip effect: The impact of forecasting, lead times and information. Management Science (1998)
8. Snyder, L., Shen, Z.M.: Fundamentals of supply chain theory, 2nd edn. (2019)
9. Gilbert, K.: An arima supply chain model. Manag. Sci. **51**(2), 305–310 (2005)
10. Li, G., Wang, S., Yan, H., Yu, G.: Information transformation in a supply chain: a simulation study. Comput. Operat. Res. **32**(3), 707–725 (2005)
11. Li, G., Yu, G., Wang, S., Yan, H.: Bullwhip and anti-bullwhip effects in a supply chain. Int. J. Product. Res. **55**(18), 5423–5434 (2017)
12. Forrester, J.: Industrial dynamics: a major breakthrough for decision makers. Harvard Bus. Rev. **36**(4), 37–66 (1958)
13. Wikner, J., Towill, D., Naim, M.: Smoothing supply chain dynamics. Int. J. Product. Econom. **22**(3), 231–248 (1991)
14. Gaalman, G., Disney, S., Wang, X.: When the bullwhip effect is an increasing function of the lead time. pp. 2297–2302 (2019)
15. Kim, I., Springer, M.: Measuring endogenous supply chain volatility: beyond the bullwhip effect. Europ. J. Operat. Res. **189**(1), 172–193 (2008)
16. Chen, F., Ryan, J., Simchi-Levi, D.: The impact of exponential smoothing forecasts on the bullwhip effect. Naval Res. Logist. **47**(4), 269–286 (2000)
17. Bayraktar, E., Lenny Koh, S., Gunasekaran, A., Sari, K., Tatoglu, E.: The role of forecasting on bullwhip effect for e-scm applications. Int. J. Product. Econ. **113**(1), 193–204 (2008)
18. Cho, D., Lee, Y.: Bullwhip effect measure in a seasonal supply chain. J. Intell. Manuf. **23**(6), 2295–2305 (2012)
19. Cho, D., Lee, Y.: The value of information sharing in a supply chain with a seasonal demand process. Comput. Ind. Eng. **65**(1), 97–108 (2013)
20. Nagaraja, C., Thavaneswaran, A., Appadoo, S.: Measuring the bullwhip effect for supply chains with seasonal demand components. Europ. J. Operat. Res. **242**(2), 445–454 (2015)
21. Najafi, M., Farahani, R.Z.: New forecasting insights on the bullwhip effect in a supply chain. IMA J Manag Math **25**(3), 259–286 (2014)
22. Rostami-Tabar, B., Babai, M., Ali, M., Boylan, J.: The impact of temporal aggregation on supply chains with arma(1,1) demand processes. Europ J Operat Res **273**(3), 920–932 (2019)
23. Jin, Y., Williams, B., Waller, M., Hofer, A.: Masking the bullwhip effect in retail: the influence of data aggregation. Int. J. Phys. Distribut. Log. Manag. **45**(8), 814–830 (2015)
24. Disney, S., Farasyn, I., Lambrecht, M., Towill, D., de Velde, W.: Taming the bullwhip effect whilst watching customer service in a single supply chain echelon. Europ. J. Operat. Res. **173**(1), 151–172 (2006)
25. Gaalman, G., Disney, S.: On bullwhip in a family of order-up-to policies with arma(2,2) demand and arbitrary lead-times. Int. J. Product. Econ. **121**(2), 454–463 (2009)
26. Wright, D., Yuan, X.: Mitigating the bullwhip effect by ordering policies and forecasting methods. Int. J. Product. Econ. **113**(2), 587–597 (2008)
27. Hosoda, T., Disney, S.: On variance amplification in a three-echelon supply chain with minimum mean square error forecasting. Omega **34**(4), 344–358 (2006)
28. Hu, Q.: Bullwhip effect in a supply chain model with multiple delivery delays. Operat. Res. Lett. **47**(1), 36–40 (2019)
29. Wang, Z., Wang, X., Ouyang, Y.: Bounded growth of the bullwhip effect under a class of nonlinear ordering policies. Europ. J. Operat. Res. **247**(1), 72–82 (2015)
30. Chatfield, D.: Underestimating the bullwhip effect: a simulation study of the decomposability assumption. Int. J. Product. Res. **51**(1), 230–244 (2013)
31. Disney, S., Naim, M., Potter, A.: Assessing the impact of e-business on supply chain dynamics. Int. J. Product. Res. **89**(2), 109–118 (2004)

32. Holmström, J.: Business process innovation in the supply chain—a case study of implementing vendor managed inventory. Europ. J. Purchas. Supply Manag. **4**(2–3), 127–131 (1998)
33. Fransoo, J., Wouters, M.: Measuring the bullwhip effect in the supply chain. Supply Chain Manag. **5**(2), 78–89 (2000)
34. Kaipia, R., Holmström, J., Tanskanen, K.: Vmi: What are you losing if you let your customer place orders. Working Paper (2000)
35. Dejonckheere, J., Disney, S., Lambrecht, M., Towill, D.: Measuring and avoiding the bullwhip effect: a control theoretic approach. Europ. J. Operat. Res. **147**(3), 567–590 (2003)
36. Chatfield, D., Kim, J., Harrison, T., Hayya, J.: The bullwhip effect—impact of stochastic lead time, information quality, and information sharing: a simulation study. Product. Operat. Manag. **13**(4), 340–353 (2004)
37. O'Donnell, T., Humphreys, P., McIvor, R., Maguire, L.: Reducing the negative effects of sales promotions in supply chains using genetic algorithms. Expert Syst. Appl. **36**(4), 7827–7837 (2009)
38. Barlas, Y., Gunduz, B.: Demand forecasting and sharing strategies to reduce fluctuations and the bullwhip effect in supply chains. J. Operat. Res. Soc. **62**(3), 458–473 (2011)
39. Shaban, A., Shalaby, M.A.: Modeling and optimizing of variance amplification in supply chain using response surface methodology. Comput. Ind. Eng. **120**, 392–400 (2018)
40. Priore, P., Ponte, B., Rosillo, R., de la Fuente, D.: Applying machine learning to the dynamic selection of replenishment policies in fast-changing supply chain environments. Int. J. Product. Res. **57**(11), 3663–3677 (2019)
41. Garg, A., Singh, S., Gao, L., Meijuan, X., Tan, C.: Multi-objective optimisation framework of genetic programming for investigation of bullwhip effect and net stock amplification for three-stage supply chain systems. Int. J. Bio-Insp. Comput. **16**(4), 241–251 (2020)
42. Alabdulkarim, A.: Minimizing the bullwhip effect in a supply chain: a simulation approach using the beer game. Simulation **96**(9), 737–752 (2020)
43. Al-Khazraji, H., Cole, C., Guo, W.: Optimization and simulation of dynamic performance of production-inventory systems with multivariable controls. Mathematics **9**(5) (2021)

Chapter 5
The Effect of the Quality of Information in SCD

5.1 Introduction

As we have seen from the previous chapter, the manner in which information flows across the SC has an enormous impact on its performance. In this chapter, we will have a closer look at this issue by examining the consequences of common problems related to this information flow, as well as the advantages that can be derived from the initiatives aimed to improve it.

As we already know from the previous chapter, a customer passes demand information to its provider in the form of orders, and the latter uses this demand information together with inventory information to place an order to its provider. Therefore, we would like to investigate how the node uses this demand and inventory information to make decisions, and what are the consequences in terms of SC dynamics if this information is not accurate, or it is not actual. Furthermore, we discuss the effect in the dynamics of the system if the demand information is communicated by the customer to the retailer in advance of the actual orders, so the node has advanced information regarding the future demand at the time of making replenishment decisions. Finally, we also know from the previous chapter that the process of multiple forecasts that takes place in the SC leads to the amplification of the variance of orders, and that it has been suggested that a remedy is to share the final customer demand data across the SC (info sharing). In this chapter, we will also discuss and quantify these advantages.

More specifically, in this chapter, we address the following issues:

- What is the effect in terms of SC performance if the final customer demand information is shared across the SC (Sect. 5.2).
- What happens if the demand data used to carry out the forecast (and the subsequent replenishment decision) are not actual (Sect. 5.3).
- What happens if the physical available inventory in the node does not match the inventory recorded in the node information system, a situation denoted as IRI (Inventory Record Inaccuracy) (Sect. 5.4).

J. M. Framinan, *Modelling Supply Chain Dynamics*,
https://doi.org/10.1007/978-3-030-79189-6_5

- How SC performance can be improved by using Advance Demand Information (ADI) (Sect. 5.5).

The picture that emerges from the conclusions in this chapter is that, while some initiatives (such as ADI) are clearly beneficial and some common problems (such as IRI) always worsen the SC performance, in some other cases related to demand data, the effect is unclear and it is linked to the demand model that can be used. Also, it is to note that, some of the models developed in this chapter are quite involving, so in these cases, guidelines showing the main rationale of the proofs have been provided for the readership that might not want to get stuck into the mathematics behind these models.

5.2 Information Sharing

Information sharing has been identified as one of the most important avenues for the reduction of the bullwhip effect. Along this line, a number of practices such as Vendor Managed Inventory (VMI), Efficient Consumer Response (ECR), Continuous Replenishment (CR) and Electronic Data Interchange (EDI) are commonplace in the industry in order to improve the visibility and transparency of the information among nodes. In this regard, there are substantial empirical evidences of the superior performance of SC implementing some type of information exchange. However, as we will see, these potential advantages are heavily influenced by a number of factors, such as the final customer demand pattern, the manner in which the nodes forecast their demand and how they use their demand data. In other words, there are situations in which information sharing does not pay off. To do so, we will reproduce some of the main findings in the academic literature regarding information sharing using models.

More specifically, in this section, we try to model a two-echelon SC such as the one depicted in Fig. 5.1. More specifically, in our model, the final demand—modelled as an AR(1) series as usual in our models—is met by a retailer, who, in turn, obtains the product from a supplier. The supplier obtains the products from a manufacturer, but this node is so far excluded from the model. Both retailer and supplier operate under a MTS policy, thus attempting to meet the corresponding demand from the

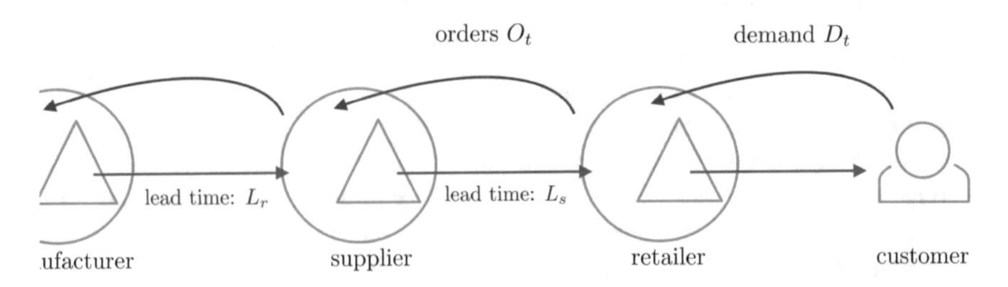

Fig. 5.1 Illustration of equivalence of forecasts

inventory. To do so, each period, they take order replenishment decisions following an OUT policy. The orders placed by the retailer (supplier) are served by the supplier (manufacturer) after L_r (L_s) time periods. Furthermore, in this section, we will assume that $\rho \geq 0$, since, as we have discussed in Chap. 2, it has been observed that the correlation is positive for most of the products.

We have already analysed, in Chap. 4, how ordering decisions are taken by the retailer. More specifically, we know from this chapter that, in each period, the retailer places the following order O_t to the supplier, see Eq. (4.24):

$$O_t = (1 + K) \cdot D_t - K \cdot D_{t-1} \tag{5.1}$$

with $K = \rho \frac{1-\rho^{L_r}}{1-\rho}$. Note that this equation is the same than Eq. (4.24), being the only difference that the retailer lead time (the only lead time considered in the models in Chap. 4) was denoted simply as L, whereas it is now denoted as L_r.

The idea now is to understand how the supplier makes ordering decisions in view of the demand from the retailer—which is made of the orders in Eq. (5.1)—and to study if having some knowledge regarding the final customer demand would impact in their decisions and on the dynamics of the SC. More specifically, at time t, we know that the supplier will make a replenishment decision in order to anticipate the retailers' orders that are expected to arrive in the next L_s periods. If he/she uses an OUT policy, then s_t^s the optimal target stock level of the supplier would be (see Sect. 2.55 in Chap. 2)

$$s_t^s = M_t + z\sqrt{V_t} \tag{5.2}$$

where M_t is the expected value of the orders received from the retailer across L_s periods, and V_t is the variance of the orders received from the retailer across L_s periods. z, as usual, is the safety factor that depends on the cost structure and/or the desired fill rate; see Sect. 2.4. In other words, Eq. (5.2) simply reflects that the target stock must face the expected demand during the lead time plus the safety stock to protect against the uncertainty in the demand. If we denote by $O_t^{L_s}$ the sum of the orders to be received by the supplier (from the retailer) in periods $t + 1, \ldots, L_s$, then

$$M_t = E[O_t^{L_s}] \tag{5.3}$$

and

$$V_t = V[O_t^{L_s}] \tag{5.4}$$

Clearly, the orders received by the supplier (placed by the retailer) contain some *implicit* information regarding the final customer demand. To make one extreme example, if the orders received by the supplier are zero during several consecutive periods, he/she might infer that the final customer demand is zero, or at least very low. However, this is only a guess, since there might be also other possible reasons making the retailer not to place orders (such as, e.g. removing some excess inventory that has been accumulated due to reasons unknown to the supplier). If the supplier could have the actual information of the customer demand, then the guess can be confirmed or refuted, and in such case, the supplier would have the *explicit* information on the customer demand.

As a result, we can see that, intuitively, there are differences if the supplier has information on the final customer demand (*information sharing*), or not (*no info sharing*). The question is, if having this information means that the average inventory of the supplier would be lower than in the case that this information is not available. As we know from Sect. 2.4.2 in Chap. 2, the average (on-hand) inventory depends on the optimal target stock in Eq. (5.2) according to the approximation in Eq. (2.60). Therefore, the average supplier inventory is potentially affected by having (or not) this information. If this is the case, we could prove that sharing final customer demand information (e.g. from POS data) would serve to reduce the bullwhip effect. Alternatively, it might be that the implicit customer demand information already contained in the retailer's orders is sufficient and there is no need to share such information across the nodes of the SC.

This section is devoted to answering the questions formulated in the above paragraph. The proofs are quite involving and, although the results are undoubtedly interesting, they require some guidance in order not to get lost with the mathematics. The program that we adopt is the following:

1. We write O_t the orders received by the supplier (from the retailer) in terms of D_t the demand from the final customer. In this manner, we embed the implicit information regarding the final customer in O_t. This is done in Sect. 5.2.1.
2. We find an expression for $O_t^{L_s}$ the total orders to be received by the supplier in the next L_s periods. Clearly, $O_t^{L_s} = \sum_{k=1}^{L_s} O_{t+k}$, so using the expressions in the first step, we will embed the implicit information on D_t in $O_t^{L_s}$. This is done in Sect. 5.2.2.
3. We find an expression for the mean and variance of $O_t^{L_s}$ in the case that there is no information sharing. This is done in Sect. 5.2.3 and we will denote by M_t^{NS} and V_t^{NS} the corresponding variables.
4. We find an expression for the mean and variance of $O_t^{L_s}$ in the case that there is information sharing. This is done in Sect. 5.2.4 and we will denote by M_t^{IS} and V_t^{IS} the corresponding variables.

Finally, we will compare and discuss the expressions of the average inventory under the info sharing and no info sharing scenarios in Sect. 5.2.5.

5.2.1 Expressing the Orders as a Function of the Customer Demand

To do so, first we try to express the demand seen by the supplier (i.e. the orders placed by the retailer) in terms of the current final demand. Taking Eq. (5.1), we see that we can write

$$O_t = D_t + K \cdot (D_t - D_{t-1}) = D_t + \rho \frac{1-\rho^{L_r}}{1-\rho} \cdot (D_t - D_{t-1}) \tag{5.5}$$

On the other hand, since we have assumed that the demand of the product can be modelled as an AR(1) time series ($D_t = d + \rho D_{t-1} + \epsilon_t$), then we can write the difference between D_t and D_{t-1} as follows:

$$D_t - \rho D_t = d + \rho(D_{t-1} - D_t) + \epsilon_t \tag{5.6}$$

and then we can manipulate the previous equation to obtain

$$\frac{1}{\rho}(D_t - \rho D_t - d - \epsilon_t) = D_{t-1} - D_t \tag{5.7}$$

or

$$D_t - D_{t-1} = \frac{1}{\rho}(d + \epsilon_t - (1-\rho)D_t) \tag{5.8}$$

Plugging the difference of the demands into Eq. (5.5), we have

$$O_t = D_t + \frac{1-\rho^{L_r}}{1-\rho}(d + \epsilon_t - (1-\rho)D_t) = \tag{5.9}$$

$$\frac{1-\rho^{L_r}}{1-\rho}(d + \epsilon_t) + \rho^{L_r} \cdot D_t \tag{5.10}$$

Before advancing, recall that what we have done so far is to write the orders placed by the retailer as a function of the demand of the final customer. In other words, we have expressed the demand perceived by the supplier as a function of the demand perceived by the retailer.

Since we would like to find an expression not only for O_t, but also for $O_{t+1}, \ldots, O_{t+L_s}$, we will try to generalise Eq. (5.9). In this regard, if we take Eq. (5.9) and write it for $t+1$, we have

$$O_{t+1} = \frac{1-\rho^{L_r}}{1-\rho}(d + \epsilon_{t+1}) + \rho^{L_r} \cdot D_{t+1} \tag{5.11}$$

Then, expressing D_{t+1} as $d + \rho D_t + \epsilon_{t+1}$ according to the AR(1) model

$$O_{t+1} = \frac{1-\rho^{L_r}}{1-\rho}(d + \epsilon_{t+1}) + \rho^{L^r} \cdot (d + \rho D_t + \epsilon_{t+1}) = \tag{5.12}$$

$$\left(\frac{1-\rho^{L_r}}{1-\rho}+\rho^{L_r}\right)(d+\epsilon_{t+1})+\rho^{L_r+1}D_t= \tag{5.13}$$

$$\frac{1-\rho^{L_r+1}}{1-\rho}(d+\epsilon_{t+1})+\rho^{L_r+1}D_t \tag{5.14}$$

and taking into account that, according to Eq. (5.9) $\rho^{L_r}D_t$ can be written as $O_t - \frac{1-\rho^{L_r}}{1-\rho}(d+\epsilon_t)$, we have

$$O_{t+1}=\frac{1-\rho^{L_r+1}}{1-\rho}(d+\epsilon_{t+1})+\rho\cdot\left(O_t-\frac{1-\rho^{L_r}}{1-\rho}(d+\epsilon_t)\right)= \tag{5.15}$$

$$d+\rho\cdot O_t+\frac{1-\rho^{L_r+1}}{1-\rho}\epsilon_{t+1}-\rho\frac{1-\rho^{L_r}}{1-\rho}\epsilon_t \tag{5.16}$$

The repeated use of Eq. (5.15) can provide a general expression for O_{t+k}. For instance, for O_{t+2} we can write, according to Eq. (5.15)

$$O_{t+2}=d+\rho\cdot O_{t+1}+\frac{1-\rho^{L_r+1}}{1-\rho}\epsilon_{t+2}-\rho\frac{1-\rho^{L_r}}{1-\rho}\epsilon_{t+1} \tag{5.17}$$

The expression in Eq. (5.15) can be plugged to substitute O_{t+1}:

$$O_{t+2}=d+\rho\cdot\left(d+\rho\cdot O_t+\frac{1-\rho^{L_r+1}}{1-\rho}\epsilon_{t+1}-\rho\frac{1-\rho^{L_r}}{1-\rho}\epsilon_t\right)+ \tag{5.18}$$

$$\frac{1-\rho^{L_r+1}}{1-\rho}\epsilon_{t+2}-\rho\frac{1-\rho^{L_r}}{1-\rho}\epsilon_{t+1}=d(1+\rho)+\rho^2\cdot O_t+\frac{1-\rho^{L_r+1}}{1-\rho}\epsilon_{t+2}+ \tag{5.19}$$

$$\left(\rho\frac{1-\rho^{L_r+1}}{1-\rho}-\rho\frac{1-\rho^{L_r}}{1-\rho}\right)\epsilon_{t+1}-\rho^2\frac{1-\rho^{L_r}}{1-\rho}\epsilon_t= \tag{5.20}$$

$$d(1+\rho)+\rho^2\cdot O_t+\frac{1-\rho^{L_r+1}}{1-\rho}\epsilon_{t+2}+\rho^{L_r+1}\epsilon_{t+1}-\rho^2\frac{1-\rho^{L_r}}{1-\rho}\epsilon_t \tag{5.21}$$

It can be seen that the general expression for O_{t+k} is

$$O_{t+k}=d\frac{1-\rho^k}{1-\rho}+\rho^k\cdot O_t+\frac{1-\rho^{L_r+1}}{1-\rho}\epsilon_{t+k}+ \tag{5.22}$$

$$\sum_{j=1}^{k-1}\rho^{L_r+j}\epsilon_{t+k-j}-\rho^k\frac{1-\rho^{L_r}}{1-\rho}\epsilon_t \tag{5.23}$$

As we can see, Eq. (5.22) gives the expression of the orders received by the supplier as a function of the customer demand. Note that, according to the sequence of events adopted in the model, at time t when the supplier would make a replenishment decision, O_t the order placed by the supplier is known (in the same manner that, when the retailer makes a replenishment decision at time t, the demand D_t is known).

5.2.2 Expressing the Orders Across the Supplier Lead Time as a Function of the Customer Demand

As discussed before, $O_t^{L_s}$ the demand that the supplier has to meet during the supplier lead time L_s is

$$O_t^{L_s} = \sum_{k=1}^{L_s} O_{t+k} \tag{5.24}$$

This expression can be computed substituting the general equation for O_{t+k}, i.e. Eq. (5.22)

$$O_t^{L_s} = \sum_{k=1}^{L_s} \left(\frac{1-\rho^k}{1-\rho} + \rho^k \cdot O_t + \frac{1-\rho^{L_r+1}}{1-\rho} \epsilon_{t+k} + \right. \tag{5.25}$$

$$\left. \sum_{j=1}^{k-1} \rho^{L_r+j} \epsilon_{t+k-j} - \rho^k \frac{1-\rho^{L_r}}{1-\rho} \epsilon_t \right) = \tag{5.26}$$

$$\frac{d}{1-\rho} \left(L_s - \rho \frac{1-\rho^{L_s}}{1-\rho} \right) + \rho \frac{1-\rho^{L_s}}{1-\rho} O_t - \rho \frac{(1-\rho^{L_r})(1-\rho^{L_s})}{(1-\rho)^2} \epsilon_t + \tag{5.27}$$

$$\sum_{k=1}^{L_s} \left(\frac{1-\rho^{L_r+1}}{1-\rho} \epsilon_{t+k} + \sum_{j=1}^{k-1} \rho^{L_r+j} \epsilon_{t+k-j} \right) \tag{5.28}$$

It can be seen that the last row in the above equation has some terms referring to the same period of the white noise, therefore, after some algebraic manipulation, we obtain

$$O_t^{L_s} = \frac{d}{1-\rho} \left(L_s - \rho \frac{1-\rho^{L_s}}{1-\rho} \right) + \rho \frac{1-\rho^{L_s}}{1-\rho} O_t \tag{5.29}$$

$$-\rho \frac{(1-\rho^{L_r})(1-\rho^{L_s})}{(1-\rho)^2} \epsilon_t + \tag{5.30}$$

$$\frac{1}{1-\rho} \sum_{k=1}^{L_s-1} \left(1-\rho^{L_s+L_r+2-k}\right) \epsilon_{t+k} + \frac{1-\rho^{L_r}}{1-\rho} \epsilon_{t+L_s} \tag{5.31}$$

Now we are prepared to analyse the mean and variance of $O_t^{L_s}$ depending on whether the supplier has information regarding the demand of the final customer.

5.2.3 Orders Average and Standard Deviation for the No Info Sharing Scenario

In the case that there is no information sharing, the supplier does not know about ϵ_t, as this is the white noise of the demand that is known by the retailer (but not by the supplier). Therefore, ϵ_t should be treated as a RV and the mean value of $O_t^{L_s}$ for the no info sharing scenario—denoted as M_t^{NS}—is

$$M_t^{NS} = \frac{d}{1-\rho}\left(L_s - \rho\frac{1-\rho^{L_s}}{1-\rho}\right) + \rho\frac{1-\rho^{L_s}}{1-\rho}O_t \tag{5.32}$$

The above expression has been obtained taking into account that $E[\epsilon_t] = 0$ for all t and that O_t is known at the time that the supplier makes the decision, therefore, it is a constant. Note, however, that M_t^{NS} is time-dependent, which is indicated by the subscript. In other words, it could be seen as a conditional expectation when O_t is considered a constant.

In this case, the variance of $O_t^{L_s}$ for the no info sharing scenario—denoted as V_t^{NS}—is taken into account, so that the ϵ_t are iid and that its variance is σ^2

$$V^{NS} = \rho^2\frac{(1-\rho^{L_r})^2(1-\rho^{L_s})^2}{(1-\rho)^4}\sigma^2 + \tag{5.33}$$

$$\frac{\sigma^2}{(1-\rho)^2}\sum_{k=1}^{L_s-1}\left(1-\rho^{L_s+L_r+2-k}\right)^2 + \frac{(1-\rho^{L_r})^2}{(1-\rho)^2}\sigma^2 = \frac{\sigma^2}{(1-\rho)^2}\cdot \tag{5.34}$$

$$\left(\rho^2\frac{(1-\rho^{L_r})^2(1-\rho^{L_s})^2}{(1-\rho)^2} + \sum_{k=1}^{L_s-1}\left(1-\rho^{L_s+L_r+2-k}\right)^2 + (1-\rho^{L_r})^2\right) \tag{5.35}$$

As opposed to M_t^{NS}, V_t^{NS} is not time-dependent (for instance, it does not depend on O_t or ϵ_t), meaning that the variance of the supplier's demand across his/her lead time is constant, and therefore, there is no need for the subscript t. As a result, in the following, we will denote it as V^{NS}.

The expression in the last equation is rather complex, but we will only need to examine its values for certain values of ρ. To do so, it is interesting to rewrite the first term in the last row, i.e.

$$\rho^2\frac{(1-\rho^{L_r})^2(1-\rho^{L_s})^2}{(1-\rho)^2} = \rho^2(1-\rho^{L_r})^2\left(\frac{1-\rho^{L_s}}{1-\rho}\right)^2 = \tag{5.36}$$

$$\rho^2(1-\rho^{L_r})^2\left(\sum_{k=1}^{L_s-1}\rho^k\right)^2 \tag{5.37}$$

so V^{NS} can be written as

$$V^{NS} = \frac{\sigma^2}{(1-\rho)^2} \cdot \tag{5.38}$$

$$\left(\rho^2(1-\rho^{L_r})^2\left(\sum_{k=1}^{L_s-1}\rho^k\right)^2 + \sum_{k=1}^{L_s-1}\left(1-\rho^{L_s+L_r+2-k}\right)^2 + (1-\rho^{L_r})^2\right) \tag{5.39}$$

5.2.4 *Orders Average and Standard Deviation for the Info Sharing Scenario*

In the case that the customer demand information is shared, the supplier knows the true value of ϵ_t at the time that he/she makes the decision, therefore, ϵ_t is no longer a RV, but a constant. In this situation, the average value of $O_t^{L_s}$ for the info sharing scenario—denoted as M_t^{IS}—is

$$M^{IS} = \frac{d}{1-\rho}\left(L_s - \rho\frac{1-\rho^{L_s}}{1-\rho}\right) + \rho\frac{1-\rho^{L_s}}{1-\rho}O_t \tag{5.40}$$

$$-\rho\frac{(1-\rho^{L_r})(1-\rho^{L_s})}{(1-\rho)^2}\epsilon_t \tag{5.41}$$

or, taking into account M_t^{NS} (the expected value in the case of no info sharing)

$$M_t^{IS} = M_t^{NS} - \rho\frac{(1-\rho^{L_r})(1-\rho^{L_s})}{(1-\rho)^2}\epsilon_t \tag{5.42}$$

With respect to the variance of $O_t^{L_s}$—denoted as V_t^{IS}—is

$$V_t^{\mathrm{IS}} = \frac{\sigma^2}{(1-\rho)^2}\sum_{k=1}^{L_s-1}\left(1-\rho^{L_s+L_r+2-k}\right)^2 + \frac{(1-\rho^{L_r})^2}{(1-\rho)^2}\sigma^2 = \frac{\sigma^2}{(1-\rho)^2} \cdot \tag{5.43}$$

$$\left(\sum_{k=1}^{L_s-1}\left(1-\rho^{L_s+L_r+2-k}\right)^2 + (1-\rho^{L_r})^2\right) \tag{5.44}$$

Note that, as in the case of the no info sharing scenario, V_t^{IS} does not depend on t, so it will be denoted in the following as V^{IS}. Combining the expression of V^{IS} with the case where there is no info sharing, we have

$$V^{IS} = V^{NS} - \rho^2\sigma^2\frac{(1-\rho^{L_r})^2(1-\rho^{L_s})^2}{(1-\rho)^4} \tag{5.45}$$

Before proceeding further, let us examine the special case when the demand is iid (i.e. $\rho = 0$). It is easy to see that $M_t^{IS} = M_t^{NS}$ and that $V^{IS} = V^{NS}$. In this case, we can obtain the first (and perhaps surprising) conclusion: there are no advantages in information sharing if the final customer demand is iid.

5.2.5 The Impact of Information Sharing on Inventory

In this section, we will try to quantify the impact of information sharing on the supplier's inventory. To do so, we first recall that, for the OUT policy, the (on-hand) inventory of a node can be approximated as follows (see Eq. (2.60) in Sect. 2.4.2):

$$\bar{I} \approx s_t + \frac{E[D_t]}{2} - E[D_t^L] \tag{5.46}$$

where D_t is, as usual, the demand observed by the node, and D_t^L is the total demand observed across the node's lead time L. Since, in our case, the demand observed by the node is O_t and the total demand observed across the supplier's lead time is $\sum_{k=1}^{L_s} O_{t+k}$, we can write

$$\bar{I} \approx s_t + \frac{E[O_t]}{2} - E\left[\sum_{k=1}^{L_s} O_{t+k}\right] \tag{5.47}$$

Taking expectations in Eq. (5.9), it is easy to find the value of $E[O_t]$ taking into account that $E[D_t] = \frac{d}{1-\rho}$

$$E[O_t] = E\left[\frac{1-\rho^{L_r}}{1-\rho}(d+\epsilon_t) + \rho^{L_r} \cdot D_t\right] = \frac{d}{1-\rho} \tag{5.48}$$

Similarly, it is easy to find the value of $E[\sum_{k=1}^{L_s} O_{t+k}]$

$$E\left[\sum_{k=1}^{L_s} O_{t+k}\right] = \frac{d \cdot L_s}{1-\rho} \tag{5.49}$$

On the other hand, we know that the quasi-optimal base stock level is given by $s_t = \mu_t^L + z\sigma_t^L$, where μ^L is the average demand across the node's lead time and σ^L the variance of the demand across this lead time. Therefore, we have that, if no information is shared regarding the customer demand, the base stock level for the supplier would be given by $s_t = M_t^{NS} + z\sqrt{V^{NS}}$ and the expression of the average on-hand inventory, in this case (denoted as $\bar{I}^{NS}$), is

$$\bar{I}^{NS} \approx M_t^{NS} + z\sqrt{V^{NS}} + \frac{d}{2(1-\rho)} - \frac{d \cdot L_s}{1-\rho} \tag{5.50}$$

As we see, the estimation of the average on-hand inventory depends on t due to M_t. To estimate the average (or long run over t) value of this approximation, we can substitute M_t by its expected value across t (as mentioned before, M_t can be see as the conditional expectation $E[O_t^L|O_t]$). According to the law of the total mean, we have

$$E[O^{L_s}] = E[E[O^{L_s}|O_t]] = E[M_t^{NS}] \tag{5.51}$$

and substituting the expression for M_t^{NS} from Eq. (5.32) and taking into account that, as we have seen previously that $E[O_t] = \frac{d}{1-\rho}$, we have

$$E[O^{L_s}] = E[\frac{d}{1-\rho}\left(L_s - \rho\frac{1-\rho^{L_s}}{1-\rho}\right) + \rho\frac{1-\rho^{L_s}}{1-\rho}O_t] = \tag{5.52}$$

$$\frac{d \cdot L_s}{1-\rho} \tag{5.53}$$

As a result, the long-run estimate of the supplier's average inventory if no information sharing (denoted as I^{NS}) is given by

$$I^{NS} \approx \frac{d}{2(1-\rho)} + z\sqrt{V^{NS}} \tag{5.54}$$

We can proceed in a similar way to estimate the supplier's average inventory when information is shared. In this case, the only difference is that the base stock level is now given by $s_t = M_t^{IS} + z\sqrt{V^{IS}}$ and, consequently, I^{IS} the average on-hand inventory can be approximated by

$$I^{IS} \approx M_t^{IS} + z\sqrt{V^{IS}} + \frac{d}{2(1-\rho)} - \frac{d \cdot L_s}{1-\rho} \tag{5.55}$$

Note that we can assume that z is the same as in the case before, since this value has been obtained according to the backlog costs and inventory holding costs, or by specifying a desired service level (see Eq. (2.53) in Chap. 2), which we can assume that are the same regardless the information that it is shared.

Substituting M_t^{IS} by $E[M_t^{IS}]$ in order to obtain the long-run estimate of I', we can see that (check Eq. (5.40)) $E[M_t^{IS}] = E[M_t^{NS}]$ as $E[\epsilon_t] = 0$. Therefore

$$I^{IS} \approx \frac{d}{2(1-\rho)} + z\sqrt{V^{IS}} \tag{5.56}$$

We can measure the impact of information sharing if we define ΔI as the relative increase of supplier's inventory caused by the fact that the retailer does not share customer demand, i.e.

$$\Delta I = \frac{I^{NS} - I^{IS}}{I^{NS}} \tag{5.57}$$

and, taking into account the approximations of I^{NS} and I^{IS} obtained in Eqs. (5.54) and (5.56), we have

$$\Delta I = \frac{z\sqrt{V^{NS}} - z\sqrt{V^{IS}}}{\frac{d}{2(1-\rho)} + z\sqrt{V^{NS}}} = \frac{\sqrt{V^{NS}(1-\rho)^2} - \sqrt{V^{IS}(1-\rho)^2}}{\frac{d}{2z} + \sqrt{V^{NS}(1-\rho)^2}} \tag{5.58}$$

Note that the denominator of the above expression is always strictly positive, so the sign of ΔI depends on the numerator. We will see that this sign depends on the values of ρ. Let us denote the functions $U^{NS}(\rho)$ and $U^{IS}(\rho)$ as follows:

$$U^{NS}(\rho) = \sqrt{V^{NS}(1-\rho)^2} = \tag{5.59}$$

$$\sigma \cdot \left(\rho^2(1-\rho^{L_r})^2 \left(\sum_{k=1}^{L_s-1} \rho^k \right)^2 + \sum_{k=1}^{L_s-1} \left(1 - \rho^{L_s+L_r+2-k}\right)^2 + (1-\rho^{L_r})^2 \right)^{\frac{1}{2}} \tag{5.60}$$

and

$$U^{IS}(\rho) = \sqrt{V^{IS}(1-\rho)^2} = \tag{5.61}$$

$$\sigma^2 \cdot \left(\sum_{k=1}^{L_s-1} \left(1 - \rho^{L_s+L_r+2-k}\right)^2 + (1-\rho^{L_r})^2 \right)^{\frac{1}{2}} \tag{5.62}$$

Clearly, the sign of ΔI depends on the difference $U^{NS}(\rho) - U^{IS}(\rho)$. Despite the complexity of both functions, it is relatively easy to see the values that they take for $\rho = 0$ and $\rho = 1$ (recall that in this section we are assuming that $0 \leq \rho < 1$). In the case of $U(\rho)$, it can be seen that $U^{NS}(0) = \sigma \cdot \sqrt{L_s}$ and that $U^{NS}(1) = 0$. Regarding $U^{IS}(\rho)$, we have that $U^{IS}(0) = \sigma \cdot \sqrt{L_s}$ and that $U^{IS}(1) = 0$. Therefore, $\Delta I(0) = 0$ and $\Delta I(1) = 0$, indicating that ΔI is non-monotonic with ρ. Indeed, for the extreme cases $\rho = 0$ and $\rho = 1$, there is no advantage of sharing information in terms of average inventory.

As a conclusion, we can state that the inventory reduction ΔI resulting from sharing the information between the retailer and the supplier is a non-monotonic function with ρ. Figure 5.2 shows how ΔI evolves depending on $\rho > 0$ and several values of L_s for a given combination of the rest of the factors ($L_r = 2$, $d = 100$, $\sigma = 50$, $z = 2$). As we can see from the figure, even with a moderate variability (the standard deviation is half of the mean), for high values of correlation, the inventory reduction can be quite important. In contrast, for low values of ρ, the savings are quite modest.

Figure 5.3 shows the inventory reduction for the same values of the demand and the safety factor ($d = 100$, $\sigma = 50$, $z = 2$). In this figure, $L_s = 2$, and it is L_r the lead time that is varying from 1 to 7. By comparing this figure and Fig. 5.2 (which are depicted using the same scale), we can see that the effect of L_r is lower than that of L_s.

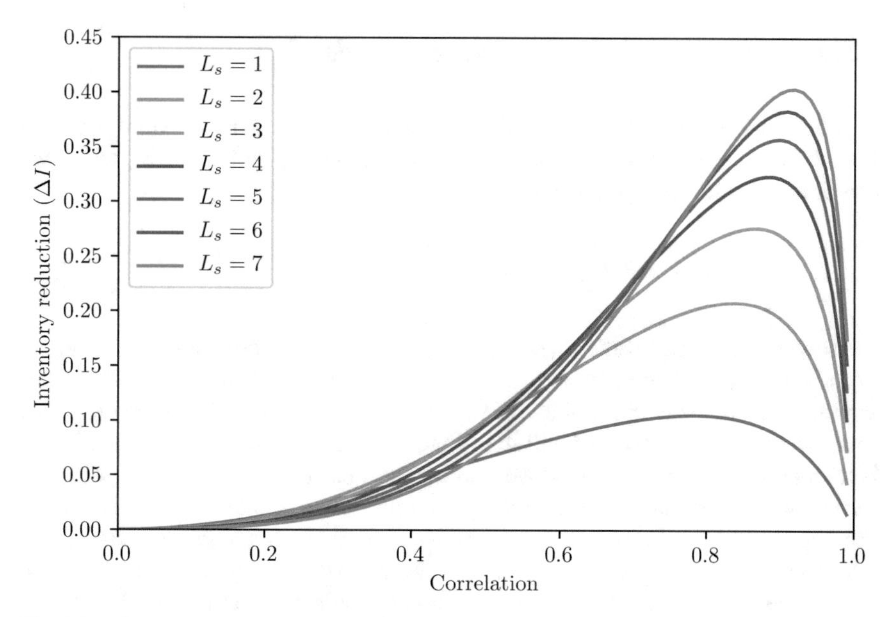

Fig. 5.2 Effect of information sharing in inventory reduction ($L_r = 2$, $d = 100$, $\sigma = 50$, $z = 2$)

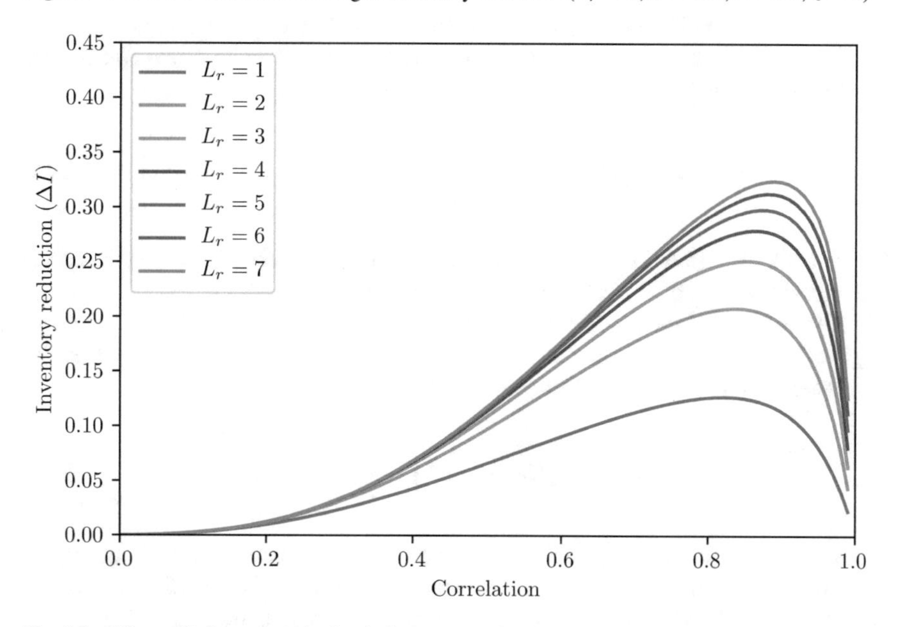

Fig. 5.3 Effect of information sharing in inventory reduction ($L_s = 2$, $d = 100$, $\sigma = 50$, $z = 2$)

5.2.6 Using Past Orders to Forecast Demand

In the model developed in the previous sections, the supplier used only the most recent order O_t to forecast future orders. This can be seen in Eq. (5.15), which we repeat here for the sake of comparison

$$O_{t+1} = d + \rho \cdot O_t + \frac{1-\rho^{L_r+1}}{1-\rho}\epsilon_{t+1} - \rho\frac{1-\rho^{L_r}}{1-\rho}\epsilon_t \tag{5.63}$$

However, also past orders $(O_{t-1}, O_{t-2}, \ldots)$ contain useful information about O_{t+1}, therefore, we will show that, if this information is used, the value of sharing customer demand diminishes. To illustrate this case, we will express O_{t+1} as a function of the past orders and will show that, in this expression, the variance of O_{t+1} is smaller than the one obtained if we use only the most recent information (Eq. 5.15). Since the equations can become quite ugly, we just give a detailed explanation for the specific case where $L_s = 1$. The conclusions that we obtain are general, something which is not surprising since, in principle, there are no specific units for L_s, so we can always interpret that we have scaled the time units to L_s to make it 1 and then L_r is also expressed in terms of L_s.

As we can see, if no information is shared and only the most recent information O_t is used, then the variance of O_{t+1} is denoted as V^{NSL} to indicate *No Sharing* (of information) and *Latest* (order) is

$$V^{\text{NSL}} = V[O_{t+1}] = \sigma^2 \cdot \left[\left(\frac{1-\rho^{L_r+1}}{1-\rho}\right)^2 + \rho^2\left(\frac{1-\rho^{L_r}}{1-\rho}\right)^2\right] \tag{5.64}$$

Note that Eq. (5.64) can be obtained also by simply making $L_s = 1$ in the corresponding equation of V^{NS} obtained in the previous section.

Alternatively, if the retailer shares the information with the supplier, ϵ_t is known by the supplier at the time that he/she makes the forecast. Therefore, in this case, the variance (denoted as V^{IS} for Information Sharing) of the order is

$$V^{\text{IS}} = V[O_{t+1}] = \sigma^2\left(\frac{1-\rho^{L_r+1}}{1-\rho}\right)^2 \tag{5.65}$$

Again, the equation is a special case for $L_s = 1$ of the generic expression of V^{IS} seen in the previous Section. By simple inspection of Eqs. (5.64) and (5.65), it is clear that $V^{\text{IS}} < V^{\text{NSL}}$ unless $\rho = 0$ (and in this case we already know that $V^{\text{IS}} = V^{\text{NSL}}$). For the sake of comparison with some expressions that will be developed below, let us rewrite Eq. (5.65) taking into account that we have usually denoted $K = \rho\frac{1-\rho_r^L}{(1-\rho)}$, and therefore, that $(1+K) = \frac{1-\rho^{L_r+1}}{(1-\rho)}$

$$V^{\text{IS}} = V[O_{t+1}] = (1+K)^2\sigma^2 \tag{5.66}$$

So far, we have just obtained the special cases of the variance of the orders in the case that information is not shared, or if it is shared, then, only information regarding the last order is used. In order to find the variance of the future orders, when all past orders are used, we start from Eq. (4.24) in Chap. 4, and try to write O_t as a function of past orders instead of a function of the current and past demand data. More specifically, we take Eq. (4.24) that expresses the order placed by the retailer at period t

$$O_t = (1+K)\cdot D_t - K\cdot D_{t-1} \tag{5.67}$$

with $K = \rho\frac{1-\rho^{L_r}}{1-\rho}$ and rewrite it

$$D_t = \frac{1}{1+K}(O_t + K\cdot D_{t-1}) \tag{5.68}$$

We can write D_{t-1} as

$$D_{t-1} = \frac{1}{1+K}(O_{t-1} + K\cdot D_{t-2}) \tag{5.69}$$

and substitute it in Eq. (5.68)

$$D_t = \frac{1}{1+K}(O_t + K\frac{1}{1+K}(O_{t-1} + K\cdot D_{t-2})) \tag{5.70}$$

$$\frac{1}{1+K}\left(O_t + \frac{K}{1+K}\cdot O_{t-1}\right) + \left(\frac{K}{1+K}\right)^2\cdot D_{t-2} \tag{5.71}$$

Using recursivity with D_{t-2}

$$D_t = \frac{1}{1+K}\left(O_t + \frac{K}{1+K}\cdot O_{t-1} + \left(\frac{K}{1+K}\right)^2\cdot O_{t-2}\right) + \tag{5.72}$$

$$\left(\frac{K}{1+K}\right)^3\cdot D_{t-3} \tag{5.73}$$

and, in general

$$D_t = \frac{1}{1+K}\sum_{j=0}^{t-1}\left(\frac{K}{1+K}\right)^j O_{t-j} + \left(\frac{K}{1+K}\right)^t\cdot D_0 \tag{5.74}$$

Now, using Eq. (4.24), we can express O_{t+1} as

$$O_{t+1} = (1+K)\cdot D_{t+1} - K\cdot D_t = \tag{5.75}$$

$$(1+K)\cdot(d+\rho\cdot D_t+\epsilon_t) - K\cdot D_t = \tag{5.76}$$

$$(1+K)d + [(1+K)\rho - K]\cdot D_t + (1+K)\epsilon_t \tag{5.77}$$

And substituting D_t by its corresponding expression in Eq. (5.74)

$$O_{t+1} = (1+K)d + \tag{5.78}$$

$$\frac{[(1+K)\rho - K]}{1+K}\left(\sum_{j=0}^{t-1}\left(\frac{K}{1+K}\right)^j O_{t-j} + \left(\frac{K}{1+K}\right)^t \cdot D_0\right) + \tag{5.79}$$

$$(1+K)\epsilon_t \tag{5.80}$$

We can obtain the variance of O_{t+1} according to this expression. We denote such variance as V^{NSA} to indicate *No Sharing* (information) and that *All* data are used. Recall that, at time t, orders $O_t, O_{t-1}, \ldots$ are known (therefore, its variance is zero), and that $V[D_t] = \frac{\sigma^2}{1-\rho^2}$

$$V^{\text{NSA}}(t) = V[O_{t+1}] = \left(\frac{[(1+K)\rho - K]}{1+K}\left(\frac{K}{1+K}\right)^t\right)^2 \frac{\sigma^2}{1-\rho^2} + \tag{5.81}$$

$$(1+K)^2\sigma^2 \tag{5.82}$$

Now, in order to compare the above expression with the variances in the order cases, we first recall that the expression depends on t, i.e. on the number of data from past orders employed. Let us assume for the moment that the series of data is sufficiently large, i.e. $t \to \infty$. In this case, it is easy to see that $\lim_{t\to\infty}\left(\frac{K}{1+K}\right)^t$ is zero since, for $0 \le \rho < 1$, we have that $\frac{K}{1+K} < 1$

$$\frac{K}{1+K} = \frac{\rho\frac{1-\rho^{L_r}}{1-\rho}}{1+\rho\frac{1-\rho^{L_r}}{1-\rho}} = \rho\frac{1-\rho^{L_r}}{1-\rho^{L_r+1}} < 1 \tag{5.83}$$

For this case, then the expression of V^{NSA} when $t\to\infty$—that we call $V^{\text{NSA}}(\infty)$—is

$$V^{\text{NSA}}(\infty) = (1+K)^2\sigma^2 = \left(1+\frac{1-\rho^{L_r}}{1-\rho}\right)^2\sigma^2 = \left(\frac{1-\rho^{L_r+1}}{1-\rho}\right)^2\sigma^2 \tag{5.84}$$

And, comparing the three expressions of the variance of the orders (V^{IS} from Eq. (5.65), V_{NSL} from Eq. (5.64) and $V^{\text{NSA}}(\infty)$ from Eq. (5.84)), we have the following relation:

$$V^{\text{NSL}} > V^{\text{NSA}}(\infty) = V^{\text{IS}} \tag{5.85}$$

This relation expresses the subtle differences of the value of information sharing: If the supplier has a sufficiently large number of data regarding the orders placed by the retailer, then there is no value in sharing demand information. This is motivated by the fact that the demand information is already contained in the orders placed by the retailer.

Furthermore, it can be shown that $V_{\text{NSL}} > V_{\text{NSA}}(t)$ for $t = 1, 2, \ldots$. As a consequence, the value of information sharing decreases with the number of periods t considered to include past information about the orders.

5.3 The Effect of Information Timeliness

In this section, we address the issue of not having actual data at the time that replenishment decisions have to be made, and how this affects the dynamic performance of the SC. More specifically, we know that, if we adopt the standard OUT policy, one of its key ingredients is to estimate the demand during the risk period or lead-time demand. If this estimation is not done based on the last available data, but on previous data (time-lagged data), it might be that, in principle, the forecast is less accurate and then the subsequent replenishment decisions are not adequate to cope with the incoming demand, which, in turn, could hinder the performance of the SC. The issue here is to assess this potential underperformance, particularly taking into account that having actual data readily available requires an IT and management infrastructure that is not certainly cost-free.

As we see, the analysis of this case is very relevant and we will obtain some interesting results. However, the mathematical formulation of the problem is relatively complex and it is easy to lose sight of the big picture. In order to avoid it, we first verbally express the program we intend to follow in the section. Later, we will develop the formulae and present the mathematical results. Finally, we will outline (verbally again) the main conclusions.

The program we intend to follow in this section is the following:

- We will find the expression for the 'best' forecast that we can perform with time-lagged data. To do so, we will use the MMSE estimate as a proxy of the best forecast that we can do, and we will develop an expression for such forecast with time-lagged data.
- We will compare this forecast with the 'best' forecasting with actual data in terms of its robustness and forecast error.
- We will derive an expression for the BWE indicator when using time-lagged data for forecast and compare it with the corresponding expression with actual data.

As we have done in most of the books, we assume that the demand can be modelled using an AR(1).

5.3.1 *Demand Estimate with Time-Lagged Data*

Up to now, we have assumed that the demand forecast made in time t was based on the last available data, i.e. demands $D_t, D_{t-1}, \ldots$. In this section, however, we will

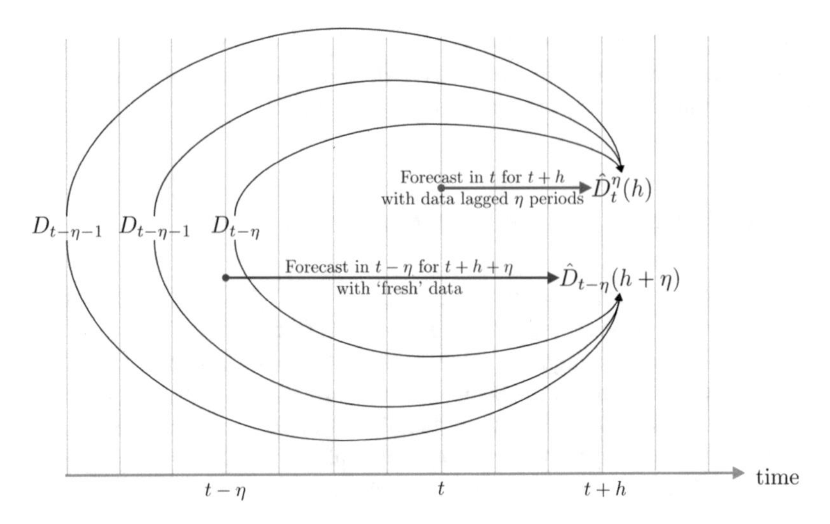

Fig. 5.4 Illustration of equivalence of forecasts

assume that, at time t, only data from period $t-\eta$ ($\eta \geq 1$) backwards are available, i.e. we only have actual demand data from periods $D_{t-\eta}, D_{t-\eta-1}, \ldots$. Let us denote by $\tilde{D}_t^\eta(h)$ this time-lagged estimate. It is easy to see that performing a forecast in period t regarding the demand in the next $t+h$ periods using data from the last $t-\eta$ periods is equivalent to perform a forecast in period $t-\eta$ (using the last available data at that time) for the next $t+h+\eta$ periods. In other words, Fig. 5.4 illustrates the equivalence, as both forecasts are built using the same data for the same time period.

As a result of the above reasoning, we have that

$$\tilde{D}_t^\eta(h) = \hat{D}_{t-\eta}(\eta+h) \tag{5.86}$$

Assuming that we perform an MMSE estimation of the demand, let us recall from Eq. (2.21) in Chap. 2 the expression for $\hat{D}_t(h)$:

$$\hat{D}_t(h) = \mu + \sum_{i=0}^{\infty} \rho^{h+i} \cdot \epsilon_{t-i}$$

We can then use the equivalence in Eq. (5.86) to determine $\tilde{D}_t^\eta(h)$

$$\tilde{D}_t^\eta(h) = \mu + \sum_{i=0}^{\infty} \rho^{h+\eta+i} \cdot \epsilon_{t-\eta-i} = \mu + \rho^{\eta+h}\epsilon_{t-\eta} + \rho^{\eta+h+1}\epsilon_{t-\eta-1} + \ldots \tag{5.87}$$

As it can be seen, some interesting conclusions can be obtained by comparing the expression of $\tilde{D}_t^\eta(h)$ (forecast with time-lagged data) with $\hat{D}_t(h)$ (forecasting with actual data): We can see that the first term in both expressions is μ followed by a series of white noise from different periods that are weighted according to ρ. In the case of $\tilde{D}_t^{\eta}(h)$, the first term in the summation is weighted by $\rho^{h+\eta}$, the second by $\rho^{h+\eta+1}$ and so forth, while in the case of $\hat{D}_t(h)$ the first term in the summation is weighted by ρ^h, the second by ρ^{h+1}, etc. Since $|\rho| < 1$, we have that in $\tilde{D}_t^{\eta}(h)$ the white noise has a lower influence than in $\hat{D}_t(h)$, and therefore, we can conclude that forecasting with time-lagged data makes the forecast to be more robust (less sensible).

Furthermore, for the special case of iid demand ($\rho = 0$), we have that both expressions are the same, so we can conclude that there is no difference between having actual data or not if these demand data are iid.

Finally and before proceeding further, let us reformulate Eq. (5.87) so that we can obtain an expression that will be useful in Section when we will assess the dynamic effects in the SC when using time-lagged data

$$\tilde{D}_t^{\eta}(h) = \mu + \sum_{i=0}^{\infty} \rho^{h+\eta+i} \cdot \epsilon_{t-\eta-i} = \mu + \rho^{h+\eta} \sum_{i=0}^{\infty} \rho^i \cdot \epsilon_{t-\eta-i} \tag{5.88}$$

If we recall the causal expression of an AR(1)—see Eq. (2.2)—we recognise that

$$D_{t-\eta} = \mu + \sum_{i=0}^{\infty} \rho^i \epsilon_{t-\eta-i}$$

so we can substitute the summation in Eq. (5.88) by $D_{t-\eta} - \mu$, and therefore, we have

$$\tilde{D}_t^{\eta}(h) = \mu + \rho^{h+\eta} \left(D_{t-\eta} - \mu\right) \tag{5.89}$$

5.3.2 The Error When Estimating with Time-Lagged Data

Now we intend to measure the error committed if we perform an MMSE estimation of the demand using the time-lagged data. In Chap. 2, we have obtained an expression for $e_t(h)$ the error of performing an MMSE estimation on a AR(1) demand with updated data—see Eq. (2.25). Given the equivalence in Eq. (5.86) with respect to using an MMSE estimation with time-lagged data, we have that the error in the latter (denoted as $\tilde{e}_t(h)$) is

$$\tilde{e}_t(h) = e_{t-\eta}(h+\eta) = \sum_{i=0}^{h+\eta-1} \rho^i \cdot \epsilon_{t+h-i} \tag{5.90}$$

We can rewrite this expression to see the relation between $\tilde{e}_t(h)$ and $e_t(h)$, i.e. the error that would be committed if the most recent data are used instead by decomposing the summation into two terms

$$\tilde{e}_t(h) = \sum_{i=0}^{h-1} \rho^i \cdot \epsilon_{t+h-i} + \sum_{i=h}^{h+\eta-1} \rho^i \cdot \epsilon_{t+h-i} = e_t(h) + \sum_{i=h}^{h+\eta-1} \rho^i \cdot \epsilon_{t+h-i}$$

Changing the index in the summation: $j = i - h$ $(i = j + h)$, we have

$$\tilde{e}_t(h) = e_t(h) + \sum_{j=0}^{\eta-1} \rho^{j+h} \cdot \epsilon_{t-j} \tag{5.91}$$

Equation (5.91) gives an interesting clue regarding the loss of accuracy of the forecast when using time-lagged data, even if the forecast is done in the best possible manner. Again, it can be seen that, for iid demand, then there is no increase in the error when using time-lagged data.

5.3.3 *The Bullwhip Effect with Time-Lagged Data*

Finally, in this section, we will obtain the value of the ratio of the variance of the orders with the variance of the demand (the BWE indicator) when time-lagged demand data are employed. Recall that, when forecasting with actual data an AR(1) demand using MMSE estimation and an OUT policy, then the expression of BWE is; see Eq. (4.26)

$$BWE = 1 + 2K(1 + K)(1 - \rho)$$

with $K = \rho\frac{1-\rho^L}{1-\rho}$.

Now we want to obtain $B\tilde{W}E$ the corresponding indicator when the forecasting (and the subsequent ordering decisions) are made with time-lagged data, i.e.

$$B\tilde{W}E = \frac{V[\tilde{O}_t]}{V[D_t]} \tag{5.92}$$

where $\tilde{O}_t$ is the order issued in period t based on time-lagged data. As we know from Chap. 4 (see Eq. 4.15), if the OUT policy is adopted, then $\tilde{O}_t$ would have the following expression:

$$\tilde{O}_t = \tilde{s}_t - \tilde{s}_{t-1} + D_t \tag{5.93}$$

where $\tilde{s}_t$ is the base stock obtained using time-lagged data

$$\tilde{s}_t = \tilde{D}_t^L + z \cdot \tilde{\sigma}_t^L \tag{5.94}$$

being $\tilde{D}_t^L$ the estimate of the demand across the risk period (using time-lagged data), and $\tilde{\sigma}_t^L$ the standard deviation of the error committed in such estimation of the demand; see Eq. (2.55) in Chap. 2.

Clearly, to compute $\tilde{O}_t$, we need to develop the expressions for $\tilde{D}_t^L$ and $\tilde{\sigma}_t^L$. Regarding the estimate of the demand across L periods, it is simply

$$\tilde{D}_t^L = \sum_{h=1}^{L} \tilde{D}_t^{\eta}(h) \tag{5.95}$$

Plugging the expression of $\tilde{D}_t^{\eta}(h)$ already obtained in Eq. (5.89), we have

$$\tilde{D}_t^L = \sum_{h=1}^{L} \left(\mu + \rho^{h+\eta}\left(D_{t-\eta} - \mu\right)\right) = \mu \cdot L + \rho^{\eta}\left(D_{t-\eta} - \mu\right)\sum_{h=1}^{L} \rho^h = \tag{5.96}$$

$$\mu \cdot L + \rho^{\eta}\left(D_{t-\eta} - \mu\right) \cdot K \tag{5.97}$$

Note that we have used K as defined above to simplify the formula.

Regarding the expression of $\tilde{\sigma}_t^L$, we will first determine the expression of $\tilde{e}_t^L$ the error committed in period t when estimating the demand across L periods, and then we obtain the standard deviation of this expression, as $\tilde{\sigma}_t^L = \sqrt{V[\tilde{e}_t^L]}$.

Clearly, this error is the sum of $\tilde{e}_t(h)$ across the L periods, i.e.

$$\tilde{e}_t^L = \sum_{h=1}^{L} \tilde{e}_t(h) \tag{5.98}$$

and, substituting the expression for $\tilde{e}_t(h)$ in Eq. (5.91), we have

$$\tilde{e}_t^L = \sum_{h=1}^{L} \left(e_t(h) + \sum_{j=0}^{\eta-1} \rho^{j+h} \cdot \epsilon_{t-j} \right) = \sum_{h=1}^{L} e_t(h) + \sum_{h=1}^{L}\sum_{j=0}^{\eta-1} \rho^{j+h} \cdot \epsilon_{t-j} \tag{5.99}$$

Note that the first term in the previous equation is simply e_t^L, the error across L periods committed when using fresh data. As we can see from Eq. (2.34), this term has the following expression:

$$e_t^L = \sum_{h=1}^{L} e_t(h) = \sum_{h=1}^{L} \frac{1}{1-\rho} \sum_{i=0}^{L-1} (1-\rho^{i+1})\epsilon_{t+L-i} \tag{5.100}$$

so if we substitute Eqs. (2.34) into (5.99), we have

$$\tilde{e}_t^L = \frac{1}{1-\rho} \sum_{h=1}^{L}\sum_{i=0}^{L-1} (1-\rho^{i+1})\epsilon_{t+L-i} + \sum_{h=1}^{L}\sum_{j=0}^{\eta-1} \rho^{j+h} \cdot \epsilon_{t-j} \tag{5.101}$$

By comparing the two summations in Eq. (5.101), we see that the first summation involves $\epsilon_{t+1}, \epsilon_{t+2}, \ldots$ while the second term involves $\epsilon_{t-\eta-1}, \ldots, \epsilon_t$. Therefore, there are no two terms of ϵ involving the same period. Taking variances in Eq. (5.101) and noting that all terms are independent

$$\left(\tilde{\sigma}_t^L\right)^2 = V[\tilde{e}_t^L] = V[e_t^L] + \sum_{h=1}^{L}\sum_{j=0}^{\eta-1} \rho^{2j}\rho^{2h} \cdot \sigma^2 = \tag{5.102}$$

$$V[e_t^L] + \rho^2 \frac{1-\rho^{2L}}{1-\rho^2}\frac{1-\rho^{2\eta}}{1-\rho^2} \cdot \sigma^2 \tag{5.103}$$

As we already know from Eq. (2.35), $V[e_t^L]$ does not depend on t, and neither does $V[\tilde{e}_t^L]$ since the second term in Eq. (5.102) does not depend on t. Therefore, $\tilde{\sigma}_t^L = \tilde{\sigma}_{t-1}^L$.

Once that we have shown that the expression of $\tilde{\sigma}_t^L$ does not depend on t, it is clear that $\tilde{O}_t$ in Eq. (5.93) can be written as

$$\tilde{O}_t = \tilde{D}_t^L - \tilde{D}_{t-1}^L + D_t \tag{5.104}$$

and, substituting the expression of $\tilde{D}_t^L$ with that obtained in Eq. (5.96), we have

$$\tilde{O}_t = D_t + \rho^\eta K(D_{t-\eta} - D_{t-\eta-1}) \tag{5.105}$$

Therefore, taking variance (recall that the covariances between D_t and D_{t-j} are $\sigma^2\rho^j$ in an AR(1)):

$$V[\tilde{O}_t] = \sigma^2\left(1 + 2\,K^2\rho^{2\eta} + 2K - 2K\rho - 2\,K^2\rho\right) \tag{5.106}$$

and computing $B\tilde{W}E$, i.e. dividing the above expression by σ^2

$$B\tilde{W}E = 1 + \rho^{2\eta}\,(2\,K(k+1)(1-\rho)) \tag{5.107}$$

Integrating the expression of BWE when actual data are used for the forecasting seen in Eq. (4.26) and presented in the beginning of this section, we have

$$B\tilde{W}E = 1 + \rho^{2\eta}\,(BWE - 1) \tag{5.108}$$

Equation 5.108 expresses the relationship between the order amplification in the case of time-lagged data as compared to that if actual data are used. Unless there is no correlation between the demand data, the bullwhip effect is always amplified, even if the coefficient of correlation is negative.

5.4 Inaccuracies in the Inventory Information (IRI)

In practice, the data available in the company's Information System (IS) regarding the number of units in the inventory does not necessarily match the actual number of units stored in the warehouse. In fact, there are several factors that may generate these discrepancies. Such discrepancies—originating from various sources, as we shall discuss—are collectively known as *Inventory Record Inaccuracies* (IRI). The sources of IRI usually considered include the following:

- *Inventory Shrinkage*. Inventory shrinkage refers to the loss of products, including theft, spoilage or damage in the inventory.
- *Product misplacement*. Product misplacement may be another cause of IRI, as some products may become temporarily unavailable due to wrong placement in the warehouse. Such misplacement makes 'invisible' part of the inventory and, such, it can be considered related to the inventory size. Note also that, in contrast to inventory shrinkage, misplaced products are eventually returned back to their intended location after inventory audits, and therefore, this phenomenon first decreases and later increases the number of units available for sale. Furthermore, they cause inventory holding costs even if the items are not available for sale.
- *Transaction errors*. Transaction errors are unintentional errors originated during inventory transactions, such as counting the inventory, or order reception (such as, e.g. scanning errors). Note that, while inventory shrinkage and product misplacement affect to the actual number of units in the warehouse, transaction errors affect only the inventory records.

The difference between these three magnitudes is (ideally) corrected when inventory audits are carried out, as misplaced products are returned back and the physical inventory and inventory record are conciliated. In the next sections, we will provide some simple models to discuss how the three sources of IRI impact the bullwhip effect. As in the basic model discussed in Chap. 4, the sequence of events in the models is the following:

1. The node receives the replenishment R_t (thus increasing the inventory). Assuming that the provider has infinite capacity, the replenishment arriving at period t corresponds to the order issued L time periods, before, i.e. $R_t = O_{t-L}$, therefore

$$I_t = I_{t-1} + O_{t-L}$$

2. The node satisfies the demand D_t. Inventories are updated according to the balance equation

$$I_t = I_{t-1} - D_t$$

Usually, Steps 1 and 2 are merged into the well-known inventory equation balance

$$I_t = I_{t-1} - D_t + O_{t-L}$$

which is more useful to derive the equations. However, for the subsequent discussion on the different types of IRI, we will retain the two steps, even if the resulting (modified) inventory balance equations will be used.

3. Work in process is updated. The goal is to have a clear idea of the current work in process, so the order issued in the previous period is also taken into account along with the replenishment just arrived. Thus

$$W_t = W_{t-1} + O_{t-1} - R_t$$

or

$$W_t = W_{t-1} + O_{t-1} - O_{t-L}$$

4. The node must issue an order to face demand in the next periods. It is assumed a target inventory level (OUT policy) s_t (how to set s_t is discussed later). Once s_t is set, an order is issued so an s_t level is reached in the inventory, taking into account both the already existing inventory I_t and the orders issued in the previous periods (work in process or W_t). Therefore

$$O_t = s_t - I_t - W_t$$

Other assumptions in the early models, such as that the backlog is permitted (I_t can be negative), or that excess returns are permitted (O_t can be negative) also hold in these models. Furthermore, we assume that the demand is iid with mean μ and standard deviation σ, and that an MMSE estimation of the demand is performed. Note also that the three sources in inventory accuracy discussed above make it necessary to distinguish among up to three types of inventory:

- Physical inventory, i.e. the number of units in the warehouse. The physical inventory at time t is denoted as I_t^p.
- Sales-available inventory, i.e. the number of inventory units that can be effectively used to satisfy demand. The difference between physical inventory and sales-available inventory is relevant in the case of product misplacement, as part of the physical inventory is not available for the customer. Available-for-sale inventory at period t will be denoted as I_t^a.
- Inventory record, i.e. the number of units in the warehouse according to the company's IS. The value of this magnitude at time t is denoted as I_t^r.

5.4.1 Shrinkage

In the case of shrinkage, both physical inventory I_t^p and available-for-sales inventory I_t^a are reduced due to damage, spoilage, etc. However, this goes unnoticed by the inventory record I_t^r, and therefore, replenishment decisions are taken considering the recorded inventory. More specifically, the sequence of events is modified as follows:

In Step 1, the node does not receive the number of units ordered $t - L$ before due to damage during transportation, theft, etc. Let us denote by Δ_t^s the number of units lost in period t due to shrinkage. Therefore, $I_t^p = I_{t-1}^p + O_{t-L} - \Delta_t^s$. The available-for-sales inventory in this case is the same, i.e. $I_t^a = I_{t-1}^a + O_{t-L} - \Delta_t^s$.

In Step 2, customer demand has to be satisfied from the available-for-sales inventory, therefore, the balance equation is as follows:

$$I_t^a = I_{t-1}^a + O_{t-L} - \Delta_t^s - D_t$$

However, for the recorded inventory, in Step 1, it is assumed that all O_{t-L} has arrived, so once the demand has been satisfied, it is assumed that all units that arrive are available for sales:

$$I_t^r = I_{t-1}^r + O_{t-L} - D_t$$

Note that this is a 'phantom' balance equation, as it only serves to update the recorded inventory in the node's information system and not to satisfy the actual demand. Clearly, the relationship between available-for-sales and recorded inventory is

$$(I_t^r - I_{t-1}^r) = (I_t^a - I_{t-1}^a) + \Delta_t^s$$

Finally, in Step 3, order size is computed on the basis of the recorded inventory, i.e.

$$O_t = s_t - I_t^r - W_t \tag{5.109}$$

Under this scenario, the difference of orders in two consecutive periods is, according to Eq. (5.109)

$$O_t - O_{t-1} = (s_t - s_{t-1}) - (I_t^r - I_{t-1}^r) - (W_t - W_{t-1})$$

or

$$O_t - O_{t-1} = (s_t - s_{t-1}) - \Delta_t^s - (W_t - W_{t-1})$$

Plugging the balance equations for W_t and I^p, we have

$$O_t = (s_t - s_{t-1}) + D_t - \Delta_t^s$$

Therefore, as it can be seen, the net effect of shrinkage in replenishment decisions is that the node demands Δ_t^s less units than those required, or, equivalently, that there is an *invisible* demand Δ_t^s dragging the inventories during each period, unless an inventory audit is conducted in time period t and, as a result, the physical and recorded inventory are reconciled. In this case, $\Delta_t^s = 0$. In order to quantify the BWE indicator, some hypotheses regarding the nature of the Δ_t^s for the time periods where the audit is not conducted must be done. Some literature argues that the shrinkage errors are mostly caused by human intervention (either in the form of errors or intentionally) and consequently, they are independent of the inventory/order size. If this is the case, we can assume that, for each time period (except that of the audit), the shrinkage takes the form of iid ϵ_t^s RVs with known mean μ_s and variance σ_s^2. As a consequence, Δ_t^s can be written as follows:

$$\Delta_t^s = \begin{cases} \epsilon_t^s & \text{if } t \bmod N \neq 0 \\ 0 & \text{otherwise} \end{cases}$$

with N the auditing cycle length, i.e. the number of time periods between consecutive audits. Therefore, the expected value of Δ_t^s can be obtained as follows:

$$E[\Delta_t^s] = E[E[\Delta_t^s|\epsilon^s]] = E\left[\sum_{i=1}^{N} \frac{\epsilon_{t-i}}{N}\right] = \frac{\mu_s(N-1)}{N}$$

Computing the variance is a bit more involving, as

$$V[\Delta_t^s] = V[E[\Delta_t^s|\epsilon^s]] + E[V[\Delta_t^s|\epsilon^s]]$$

where

$$E[\Delta_t^s|\epsilon^s] = \frac{1}{N}\sum_{i=1}^{N-1} \epsilon_{t-i}$$

and thus

$$V[E[\Delta_t^s|\epsilon^s]] = \frac{N-1}{N^2}\sigma_s^2$$

On the other hand

$$V[\Delta_t^s|\epsilon^s] = E[\left(\Delta_t^s|\epsilon^s\right)^2] - E[\Delta_t^s|\epsilon^s]^2$$

where the first term is

$$E[\left(\Delta_t^s|\epsilon^s\right)^2] = \frac{1}{N}\sum_{i=1}^{N-1}(\epsilon_{t-i}^s)^2$$

and the second term is

$$E[\Delta_t^s|\epsilon^s]^2 = \left(\sum_{i=1}^{N-1} \frac{\epsilon_{t-i}}{N}\right)^2$$

Putting together the two terms and taking expectations

$$E[V[\Delta_t^s|\epsilon^s]] = \frac{1}{N}(N-1)(\sigma_s^2 + \mu_s^2) - \frac{N-1}{N^2}(\sigma_s^2 + \mu_s^2) = \tag{5.110}$$

$$\frac{N-1}{N^2}(\sigma_s^2 + \mu_s^2)(N-1) = \left(\frac{N-1}{N}\right)^2 (\sigma_s^2 + \mu_s^2) \tag{5.111}$$

From this result, it follows that

$$V[\Delta_t^s] = \frac{N-1}{N^2}\sigma_s^2 + \left(\frac{N-1}{N}\right)^2 (\sigma_s^2 + \mu_s^2) = \tag{5.112}$$

$$\frac{N-1}{N^2}\left[(N+2)\sigma_s^2 + (N+1)\mu_s^2\right] \tag{5.113}$$

Consequently, the BWE indicator in the case of shrinkage (BWE_s) is

$$BWE_s = 1 + \frac{N-1}{N^2 \cdot \sigma^2}\left[(N+2)\sigma_s^2 + (N+1)\mu_s^2\right] \tag{5.114}$$

5.4.2 Product Misplacement

In the case of product misplacement, both the physical and the recorded inventories are not reduced, but the available-for-sales inventory is reduced.

In Step 1, the node receives the number of units ordered $t - L$, therefore, the physical inventory does not change in this type of IRI and the following (original) physical inventory balance equation (before satisfying demand) holds:

$$I_t^p = I_{t-1}^p + O_{t-L}$$

However, due to the product misplacement, the available-for-sales inventory is not the same, as in general, we assume that Δ_t^m units of the arriving product have been misplaced in period t, therefore

$$I_t^a = I_{t-1}^a + O_{t-L} - \Delta_t^m$$

In Step 2, demand is satisfied from available-for-sales inventory, therefore, the balance equation is

$$I_t^a = I_{t-1}^a + O_{t-L} - \Delta_t^m - D_t$$

Similarly, in case of shrinkage, the inventory record goes unnoticed and assumes that all units arriving are available for sales, so the recorded inventory balance equation (again a 'phantom' balance equation, as all sales cannot be satisfied from O_{t-L} due to product misplacement) is

$$I_t^r = I_{t-1}^r + O_{t-L} - D_t$$

Clearly, the relationship between available-for-sales and recorded inventory is

$$(I_t^r - I_{t-1}^r) = (I_t^a - I_{t-1}^a) + \Delta_t^m$$

In Step 3, the order size is computed using the recorded inventory, so after some substitutions as in the shrinkage case, the following equation holds:

$$O_t = (s_t - s_{t-1}) + D_t - \Delta_t^m$$

If the product misplacement is considered to be independent of the inventory/order size, it can be assumed that, for each period t where there is no audit, the inventory reduction takes the form of iid ϵ_t^m RVs with known mean $\mu_m > 0$ and variance σ_m^2. In the period where the audit is performed, all the misplaced inventory is placed correctly, therefore, it is available for sales and taken into account for replenishment decisions.

As a consequence, Δ_t^m can be written as follows:

$$\Delta_t^m = \begin{cases} \epsilon_t^m & \text{if } t \bmod N \neq 0 \\ -\sum_{i=1}^{N-1} \epsilon_{t-i}^m & \text{otherwise} \end{cases}$$

with N the auditing cycle length, i.e. the number of time periods between consecutive audits. To compute the variance of Δ_t^m, we have

$$V[\Delta_t^m] = V[E[\Delta_t^m|\epsilon^m]] + E[V[\Delta_t^m|\epsilon^m]]$$

Regarding the first term of the expression, we first need to obtain the conditioned expected value of Δ_t^m

$$E[\Delta_t^m|\epsilon^m] = \frac{1}{N}\sum_{k=1}^{N-1} \epsilon_{t-i}^m - \frac{1}{N}\sum_{i=1}^{N-1} \epsilon_{t-i}^m = 0$$

so its variance is also zero. Regarding the second term, we first compute the conditional variance of Δ_t^m

$$V[\Delta_t^m|\epsilon^m] = E[(\Delta_t^m|\epsilon^m)^2] - E[\Delta_t^m|\epsilon^m]^2$$

It is clear that the second term is zero, so operating in the first term

$$V[\Delta_t^m|\epsilon^m] = E[(\Delta_t^m|\epsilon^m)^2] = \sum_{i=1}^{N-1} \frac{1}{N}(\epsilon_{t-i}^m)^2 + \frac{1}{N}\left(-\sum_{i=1}^{N-1}\epsilon_{t-i}^m\right)\cdot\left(-\sum_{i=1}^{N-1}\epsilon_{t-i}^m\right)$$

Taking expectations (and considering that $E[\epsilon_{t-i}^m \cdot \epsilon_{t-j}^m] = 0$ unless $i = j$)

$$E[V[\Delta_t^m|\epsilon^m]] = 2\frac{N-1}{N}(\sigma_m^2 + \mu_m^2)$$

Consequently, under the hypotheses of MMSE estimation of iid demands, the bullwhip effect caused my the product misplacement BWE_m is

$$BWE_m = 1 + 2\frac{N-1}{N \cdot \sigma^2}(\sigma_m^2 + \mu_m^2) \tag{5.115}$$

5.4.3 Transaction Errors

Transaction errors do not affect to physical or available-to-sales inventory, but only the recorded inventory. As a result, the number of units registered in the node's IS is different than the actual number of units in the warehouse and, consequently, ordering/replenishment decisions are based on wrong calculations.

In Step 1, the node receives the number of units ordered $t - L$, therefore, the physical inventory does not change in this type of IRI and the following (original) physical inventory balance equation (before satisfying demand) holds

$$I_t^p = I_{t-1}^p + O_{t-L}$$

Similarly, since there is no product misplacement, in Step 1 all physical inventory is available for sales

$$I_t^a = I_{t-1}^a + O_{t-L}$$

In Step 2, demand is satisfied from available-for-sales inventory, therefore, the balance equation is the same as in the base (i.e. no IRI) case

$$I_t^a = I_{t-1}^a + O_{t-L} - D_t$$

However, in this case, the transaction errors makes the recorded inventory to be different. By denoting Δ_t^e, the number of units wrongly recorded due to the transaction errors, we have (again a 'phantom' balance equation)

$$I_t^r = I_{t-1}^r + O_{t-L} - D_t + \Delta_t^e$$

Clearly, the relationship between available-for-sales and recorded inventory is

$$(I_t^r - I_{t-1}^r) = (I_t^a - I_{t-1}^a) - \Delta_t^e$$

In Step 3, the order size is computed using the recorded inventory, so after some substitutions as in the shrinkage case, the following equation holds:

$$O_t = (s_t - s_{t-1}) + D_t + \Delta_e^m$$

If the transaction errors are considered to be independent of the inventory/order size, it can be assumed that for each period t where there is no audit, the inventory variation (reduction or increase) takes the form of iid ϵ_t^m RVs with known mean μ_e and variance σ_e^2. Furthermore, if we assume that the transaction errors are unbiased, then $\mu_e = 0$. In the period where the audit is performed, all the misplaced inventory is placed correctly, therefore, it is available for sales and taken into account for replenishment decisions.

As a consequence, Δ_t^e can be written as follows:

$$\Delta_t^e = \begin{cases} \epsilon_t^e & \text{if } t \bmod N \neq 0 \\ -\sum_{i=1}^{N-1} \epsilon_{t-i}^e & \text{otherwise} \end{cases}$$

with N the auditing cycle length, i.e. the number of time periods between consecutive audits. To compute the variance of Δ_t^e, we have

$$V[\Delta_t^e] = V[E[\Delta_t^e|\epsilon^e]] + E[V[\Delta_t^e|\epsilon^e]]$$

Regarding the first term of the expression, we first need to obtain the conditioned expected value of Δ_t^e

$$E[\Delta_t^e|\epsilon^e] = \frac{1}{N}\sum_{k=1}^{N-1} \epsilon_{t-i}^e - \frac{1}{N}\sum_{i=1}^{N-1} \epsilon_{t-i}^e = 0$$

so its variance is also zero. Regarding the second term, we first compute the conditional variance of Δ_t^e

$$V[\Delta_t^e|\epsilon^e] = E[(\Delta_t^e|\epsilon^e)^2] - E[\Delta_t^e|\epsilon^e]^2$$

It is clear that the second term is zero, so operating in the first term

$$V[\Delta_t^e|\epsilon^e] = E[(\Delta_t^e|\epsilon^e)^2] = \sum_{i=1}^{N-1} \frac{1}{N}(\epsilon_{t-i}^e)^2 + \frac{1}{N}\left(-\sum_{i=1}^{N-1} \epsilon_{t-i}^e\right) \cdot \left(-\sum_{i=1}^{N-1} \epsilon_{t-i}^e\right)$$

Taking expectations (and considering that $E[\epsilon_{t-i}^e \cdot \epsilon_{t-j}^e] = 0$ unless $i = j$)

$$E[V[\Delta_t^e|\epsilon^e]] = 2\frac{N-1}{N}(\sigma_e^2 + \mu_e^2)$$

Therefore, $V[\Delta_t^e] = 2\frac{N-1}{N}(\sigma_e^2 + \mu_e^2)$
Consequently, the bullwhip effect caused by the transaction errors BWE_e is

$$BWE_e = 1 + 2\frac{N-1}{N \cdot \sigma^2}(\sigma_e^2 + \mu_e^2) \tag{5.116}$$

Furthermore, if we assume that the transaction errors are unbiased, the expression reduces to

$$BWE_e = 1 + 2\frac{N-1}{N}\left(\frac{\sigma_e}{\sigma}\right)^2 \tag{5.117}$$

5.5 Advance Demand Information

In our case, the customer shares demand information with the retailer in advance of the order due dates. This is called ADI (Advanced Demand Information). ADI can be encountered both in make-to-order and make-to-stock settings. As we will see in the next subsections, in a fairly usual make-to-stock setting, the use of ADI is equivalent to reducing the lead time, and therefore, as a way to reduce both the order and inventory variance amplification.

5.5.1 Perfect ADI in Make-to-Stock SCs

Let us consider the case of one retailer and one supplier plus the final customer. The customer places the orders to the retailer in advance, i.e. for a demand D_t to take place in period t, the advanced order request is sent to the retailer in period D_{t-L^D}. L^D is the so-called *demand lead time*, meaning the number of periods from the order arrival until its due date. Furthermore, the time required for the supplier to replenish an order placed by the retailer is denoted as L^S or supply lead time. Finally, i is considered that the supplier has unlimited capacity.

In this situation, two cases can be discussed:

- The case where $L^D \geq L^S$, i.e. the time that the order is notified in advance to the retailer is higher than the time required to replenish an order from the supplier. Clearly, in this case, the retailer does not need to build any inventory, as she is able to fulfil the demand from the final customer simply by waiting until period $L^D - L^S$ and then placing an order to the retailer. In this manner, the replenishment would arrive just in time to be served to the final customer.
- The case where $L^D < L^S$ it is by far more compelling, as it might be interesting to know whether there is some advantage for the retailer to obtain this advance

information from the customer. In this case, it is easy to see that the effect of the advanced information is to effectively shorten the supply lead time by $L = L^S - L^D$. Intuitively, it can be seen that, in a conventional (no ADI) system, for a demand to take place in period t, the supply would arrive in period $t + L$, whereas in an ADI system, such demand would arrive in $t + L^S - L^D$.

5.5.2 *The Case of Imperfect ADI*

In the previous section, we have assumed that the information provided in advance by the final customer is completely reliable, so the retailer fully relies on it to make replenishment decisions. This is named *perfect* ADI. However, in some cases, the information may not be reliable regarding the time and/or the quantity of the orders (*imperfect* ADI). Furthermore, the ADI may not be available from all customers, but only from a portion of them. In this case, even if the information provided is not very reliable, there might be some savings with respect to ignoring the advance signal. However, it is to note that traditional base stock policies may not perform well in this setting.

5.6 Conclusions

In this chapter, we have considered different issues and practices related to the flow of information in the SC. First, the potential advantages of sharing the data with the final customer demand have been discussed in Sect. 5.2. From the models developed in this section, we can conclude that there may be some advantages when sharing information, but that these are constrained by a number of factors, both exogenous and endogenous to the SC. These are:

- The nature of the customer demand. The models presented in Sect. 5.2.5 show that these advantages are zero in the case that the demand is stationary (iid), or are severely constrained if the demand shows a high positive correlation.
- The use of past information (orders) to predict future demand. Since the final customer demand information is embedded in the orders placed by the retailer to the supplier, the use of the (past) information provided by these orders helps in reducing the disadvantages of not having the final customer demand data and, in the long run, in compensating this unavailability.
- The forecast technique employed in the SC, as in the models developed in Sect. 5.2.5, we have assumed an MMSE estimation. However, many companies rely on more simple forecasting techniques such as MA or exponential smoothing and there are contributions showing that, in these cases, there may be substantial advantages in sharing the information.

The use of demand delayed information in the SC has been discussed in Sect. 5.3. The simple model obtained shows that, unless the final demand is iid, having lagged demand information increases the bullwhip effect, and the increase is positively related to the delay. Interestingly, this also occurs in the case of a negative correlation of the demand. In the model developed, we have assumed that we use the classical OUT policy. However, it might be possible to use different replenishment policies to mitigate the negative effects on the delayed information. Finally, it will be interesting to extend this simple model to more nodes in the SC in order to quantify the advantages of using actual information in the SC consisting of additional nodes. As some authors have shown, these advantages are quickly reduced for the upstream nodes in the SC.

Inventory Record Inaccuracy (IRI) has been discussed in Sect. 5.4. The three classical sources of IRI have been discussed and modelled using relatively simple hypotheses. In these models (developed for the case of iid demand), we can see that IRI increases the bullwhip effect and, although the impact is different depending on the source, this effect is proportional to the auditing cycle length. The strategies to cope with IRI may thus range from reducing or removing the root causes via the implementation of technical and/or process improvements, such as RFID (Radio Frequency Identification) using adequate inventory auditing policies that may serve to fix the discrepancies. In any case, the potential advantages of IRI reduction might have to be weighed against the additional costs incurred during the implementation of the technical/process improvements. The models discussed in the previous sections may give some estimates of this trade-off.

Finally, we have seen that the main effect of ADI is to effectively reduce the lead times as long as this advanced information is perfect. If this is not the case, there might be some advantages, but these have to be carefully quantified also taking into account that the usual OUT policy adopted in most of the models is not optimal in this case.

5.7 Further Readings

The quality of the information in supply chain dynamics is among the most important topics in the field, given its huge impact on the supply chain performance. There are a number of case studies and empirical evidence showing the value of using information to collaborate in order to increase SC performance, including [1–4]. The pioneering reference dealing with the advantages of information sharing (upon which the models show in Sect. 5.2 are based) is [5]. However, their proof had a flaw that was detected in [6], who show that the advantages of on-hand inventory are non-monotonic with the demand correlation. The use of all past order data as compared to using only the last available data is discussed in [7]. In a similar vein, the work by [8] reviews the then existing works on info sharing and reaches similar conclusions in terms of the limits of information sharing. In [2], the differences observed between the limited advantages of information sharing shown in the models discussed in Sect. 5.2 and the tangible benefits encountered in practice are discussed and reconciled. A model to discuss info sharing with generalised demand is presented in [9]. Using a simulation model with variable lead times, [10] show that info sharing does not remove the bullwhip effect in a serial SC, even if this effect is *decelerated*. These limits of info sharing are also exposed in [11], as the authors show that there are certain cases where the value of information is negative. References dealing with information sharing when the parameters of the demand process have to be estimated are [12, 13]. The effect of info sharing in the cost reduction has been studied by [14], and in the case of online retailers by [15]. The impact of different ordering policies in the info sharing context is addressed by [16]. Several studies have been carried out on information sharing in multi-level SCs using continuous, simulation, such as [17–19], using discrete-event simulation models, such as [20, 21], or with a control systems approach [22]. Sharing information in complex SC structures has also been studied in some contributions, such as in [23, 24]. Although usually considered a yes/no decision, partial/asymmetric information sharing has been also studied in the literature; see e.g. [25–28]. Finally, there are papers addressing behavioural aspects of information sharing in SCs, such as [29–31], or its role on SC resilience [32]. The impact of information sharing in the company's performance in terms of order fulfilment and its connection to that in terms of bullwhip effect is addressed by [33].

The classical reference for the effect of time-lagged data in the dynamics of the SC is [34]. In [35], a generalization of the OUT policy is proposed in order to overcome the disadvantages of the information delay identified in Sect. 5.3, whereas in [36], it is shown that the benefits of having small delays (or no delays) in the demand data employed are quite reduced for the upstream nodes in the SC, and therefore, they might be less inclined to reduce these delays.

Among the first ones to study IRI in the supply chain context are [37]. The three sources of IRI discussed in Sect. 5.4 are taken from [38]. An empirical study by [39] shows that the inventory discrepancies can be positive or negative. The strategies to deal with IRI are described in [40]. The models discussed in Sect. 5.4 are taken from [41]. Regarding shrinkage—which has been by the far most studied type of IRI—in [42], it is argued that shrinkage errors are mostly caused by human intervention (either in the form of errors or intentionally) and consequently, they are independent of the inventory/order size. Other references with different hypotheses—i.e. the shrinkage error is related to the order level or to the inventory level—are [43, 44], respectively. The advantages of using technologies such as RFID to reduce shrinkage errors are discussed in [45–47]. Other papers discussing shrinkage are [48–50].

In [51], it is shown that perfect ADI improves the performance of the system in the same manner as a reduction in the lead times. An state-dependent OUT policy is shown to be optimal in this case by [52]. Different variants of ADI (including the perfect and imperfect versions that have been discussed in Sect. 5.5) are analysed in [53]. Other papers dealing with ADI in the context of SC are [54–58].

References

1. Boone, T., Ganeshan, R.: The value of information sharing in the retail supply chain: two case studies. Foresight: Int. J. Appl. Forecast. **9**, 12–17 (2008)
2. Hosoda, T., Naim, M., Disney, S., Potter, A.: Is there a benefit to sharing market sales information? linking theory and practice. Comput. Ind. Eng. **54**(2), 315–326 (2008)
3. Kulp, S., Lee, H., Ofek, E.: Manufacturer benefits from information integration with retail customers. Manag. Sci. **50**(4), 431–444 (2004)
4. Zhou, H., Benton Jr., W.: Supply chain practice and information sharing. J. Oper. Manag. **25**(6), 1348–1365 (2007)
5. Lee, H., So, K., Tang, C.: Value of information sharing in a two-level supply chain. Manag. Sci. **46**(5), 626–643 (2000)
6. Babai, M.Z., Boylan, J.E., Syntetos, A.A., Ali, M.M.: Reduction of the value of information sharing as demand becomes strongly auto-correlated. Int. J. Prod. Econ. **181**, 130–135 (2016)
7. Raghunathan, S.: Information sharing in a supply chain: a note on its value when demand is nonstationary. Manag. Sci. **47**(4), 605–610 (2001)
8. Agrawal, S., Sengupta, R., Shanker, K.: Impact of information sharing and lead time on bullwhip effect and on-hand inventory. Eur. J. Oper. Res. **192**(2), 576–593 (2009)
9. Chen, L., Lee, H.: Information sharing and order variability control under a generalized demand model. Manag. Sci. **55**(5), 781–797 (2009)
10. Chatfield, D., Kim, J., Harrison, T., Hayya, J.: The bullwhip effect—impact of stochastic lead time, information quality, and information sharing: a simulation study. Prod. Oper. Manag. **13**(4), 340–353 (2004)
11. Teunter, R., Babai, M., Bokhorst, J., Syntetos, A.: Revisiting the value of information sharing in two-stage supply chains. Eur. J. Oper. Res. **270**(3), 1044–1052 (2018)
12. Hosoda, T., Disney, S.: Impact of market demand mis-specification on a two-level supply chain. Int. J. Prod. Econ. **121**(2), 739–751 (2009)
13. Pastore, E., Alfieri, A., Zotteri, G., Boylan, J.: The impact of demand parameter uncertainty on the bullwhip effect. Eur. J. Oper. Res. **283**(1), 94–107 (2020)
14. Cachon, G., Fisher, M.: Supply chain inventory management and the value of shared information. Manag. Sci. **46**(8), 1032–1048 (2000)
15. Zhao, J., Zhu, H., Zheng, S.: What is the value of an online retailer sharing demand forecast information? Soft Comput. **22**(16), 5419–5428 (2018)
16. Costantino, F., Di Gravio, G., Shaban, A., Tronci, M.: The impact of information sharing on ordering policies to improve supply chain performances. Comput. Ind. Eng. **82**, 127–142 (2015)
17. Fiala, P.: Information sharing in supply chains. Omega **33**(5), 419–423 (2005)
18. Cannella, S., Ciancimino, E., Framinan, J.: Inventory policies and information sharing in multi-echelon supply chains. Prod. Plan. Control **22**(7), 649–659 (2011)
19. Jeong, K., Hong, J.D.: The impact of information sharing on bullwhip effect reduction in a supply chain. J. Intell. Manuf. **30**(4), 1739–1751 (2019)
20. Kelepouris, T., Miliotis, P., Pramatari, K.: The impact of replenishment parameters and information sharing on the bullwhip effect: a computational study. Comput. Oper. Res. **35**(11), 3657–3670 (2008)
21. Hall, D., Saygin, C.: Impact of information sharing on supply chain performance. Int. J. Adv. Manuf. Technol. **58**(1–4), 397–409 (2012)
22. Ouyang, Y.: The effect of information sharing on supply chain stability and the bullwhip effect. Eur. J. Oper. Res. **182**(3), 1107–1121 (2007)
23. Dominguez, R., Framinan, J., Cannella, S.: Serial versus divergent supply chain networks: a comparative analysis of the bullwhip effect. Int. J. Prod. Res. **52**(7), 2194–2210 (2014)
24. Rached, M., Bahroun, Z.: Vertical and horizontal impacts of information sharing on a divergent supply chain. Int. J. Logist. Syst. Manag. **35**(2), 246–272 (2020)
25. Costantino, F., Di Gravio, G., Shaban, A., Tronci, M.: The impact of information sharing and inventory control coordination on supply chain performances. Comput. Ind. Eng. **76**, 292–306 (2014)

26. Dominguez, R., Cannella, S., Barbosa-Póvoa, A., Framinan, J.: Ovap: a strategy to implement partial information sharing among supply chain retailers. Transp. Res. Part E: Logist. Transp. Rev. **110**, 122–136 (2018)
27. Lei, H., Wang, J., Shao, L., Yang, H.: Ex post demand information sharing between differentiated suppliers and a common retailer. Int. J. Prod. Res. **58**(3), 703–728 (2020)
28. Dominguez, R., Cannella, S., Framinan, J.: Remanufacturing configuration in complex supply chains. Omega (United Kingdom) **101** (2021)
29. Li, C.: Controlling the bullwhip effect in a supply chain system with constrained information flows. Appl. Math. Model. **37**(4), 1897–1909 (2013)
30. Wu, I.L., Chuang, C.H., Hsu, C.H.: Information sharing and collaborative behaviors in enabling supply chain performance: a social exchange perspective. Int. J. Prod. Econ. **148**, 122–132 (2014)
31. Seifbarghy, M., Darvish, M., Akbari, F.: Analysing bullwhip effect in supply networks under information sharing and exogenous uncertainty. Int. J. Ind. Syst. Eng. **26**(3), 291–317 (2017)
32. Scholten, K., Schilder, S.: The role of collaboration in supply chain resilience. Supply Chain Manag. **20**(4), 471–484 (2015)
33. Ojha, D., Sahin, F., Shockley, J., Sridharan, S.: Is there a performance tradeoff in managing order fulfillment and the bullwhip effect in supply chains? the role of information sharing and information type. Int. J. Prod. Econ. **208**, 529–543 (2019)
34. Zhang, X.: Delayed demand information and dampened bullwhip effect. Oper. Res. Lett. **33**(3), 289–294 (2005)
35. Hosoda, T., Disney, S.M.: On the replenishment policy when the market demand information is lagged. Int. J. Prod. Econ. **135**(1), 458–467 (2012). Advances in Optimization and Design of Supply Chains
36. Hosoda, T., Disney, S.: A delayed demand supply chain: incentives for upstream players. Omega **40**(4), 478–487 (2012)
37. Kull, T., Barratt, M., Sodero, A., Rabinovich, E.: Investigating the effects of daily inventory record inaccuracy in multichannel retailing. J. Bus. Logist. **34**(3), 189–208 (2013)
38. Lee, H., Özer, Ö.: Unlocking the value of RFID. Prod. Oper. Manag. **16**(1), 40–64 (2007)
39. Ishfaq, R., Raja, U.: Empirical evaluation of IRI mitigation strategies in retail stores. J. Oper. Res. Soc. (2019)
40. Dehoratius, N., Mersereau, A., Schrage, L.: Retail inventory management when records are inaccurate. Manuf. Serv. Oper. Manag. **10**(2), 257–277 (2008)
41. Framinan, J., Cannella, S., Dominguez, R.: Investigating the effect of inventory record inaccuracy in supply chain dynamics. Tech. Rep. TROI-2020-02, Industrial Management Research Group (2020)
42. Dai, H., Tseng, M.: The impacts of rfid implementation on reducing inventory inaccuracy in a multi-stage supply chain. Int. J. Prod. Econ. **139**(2), 634–641 (2012)
43. Rekik, Y., Syntetos, A., Glock, C.: Modeling (and learning from) inventory inaccuracies in e-retailing/b2b contexts. Decis. Sci. **50**(6), 1184–1223 (2019)
44. Cannella, S., Framinan, J., Bruccoleri, M., Barbosa-Póvoa, A., Relvas, S.: The effect of inventory record inaccuracy in information exchange supply chains. Eur. J. Oper. Res. **243**(1), 120–129 (2015)
45. Alexander, K., Gilliam, T., Gramling, K., Grubelic, C., Kleinberger, H., Leng, S., Moogimane, D., Sheedy, C.: Applying auto-id to reduce losses associated with product obsolescence. Applying Auto-ID to Reduce Losses Associated with Shrink (2002)
46. Qin, W., Zhong, R., Dai, H., Zhuang, Z.: An assessment model for rfid impacts on prevention and visibility of inventory inaccuracy presence. Adv. Eng. Inform. **34**, 70–79 (2017)
47. Drakaki, M., Tzionas, P.: Investigating the impact of inventory inaccuracy on the bullwhip effect in rfid-enabled supply chains using colored petri nets. J. Model. Manag. **14**(2), 360–384 (2019)
48. Dai, H., Li, J., Yan, N., Zhou, W.: Bullwhip effect and supply chain costs with low- and high-quality information on inventory shrinkage. Eur. J. Oper. Res. **250**(2), 457–469 (2016)

49. Cannella, S., Dominguez, R., Framinan, J.: Inventory record inaccuracy—the impact of structural complexity and lead time variability. Omega (United Kingdom) **68**, 123–138 (2017)
50. Bruccoleri, M., Cannella, S., La Porta, G.: Inventory record inaccuracy in supply chains: the role of workers' behavior. Int. J. Phys. Distrib. Logist. Manag. **44**(10), 796–819 (2014)
51. Hariharan, R., Zipkin, P.: Customer-order information, lead-times, and inventories. Manag. Sci. **41**(10), 1599–1607 (1995)
52. Gallego, G., Özer, O.: Integrating replenishment decisions with advance demand information. Manag. Sci. **47**(10), 1344–1360 (2001)
53. Benbitour, M., Sahin, E.: Evaluation of the impact of uncertain advance demand information on production/inventory systems. pp. 1738–1743 (2015)
54. Chinna Pamulety, T., Pillai, V.: Performance analysis of supply chains under customer demand information sharing using role play game. Int. J. Ind. Eng. Comput. **3**(3), 337–346 (2012)
55. Tan, T., Güllü, R., Erkip, N.: Modelling imperfect advance demand information and analysis of optimal inventory policies. Eur. J. Oper. Res. **177**(2), 897–923 (2006)
56. Thonemann, U.: Improving supply-chain performance by sharing advance demand information. Eur. J. Oper. Res. **142**(1), 81–107 (2002)
57. Zhu, K., Thonemann, U.: Modeling the benefits of sharing future demand information. Oper. Res. **52**(1), 136–147 (2004)
58. Topan, E., Tan, T., van Houtum, G.J., Dekker, R.: Using imperfect advance demand information in lost-sales inventory systems with the option of returning inventory. IISE Trans. **50**(3), 246–264 (2018)

Chapter 6
Enriching SC Models

6.1 Introduction

In this chapter, the basic model presented in Chap. 4 is enriched by removing some hypotheses and including additional constraints in order to contemplate a wider range of scenarios observed in practice. One of the most relevant enhancements is related to the consideration of variable lead times (as often happen in real life), which may impact profoundly the dynamics of the supply chain. Furthermore, the variability in lead times opens the possibility of the crossover of orders, affecting the dynamics of the supply chain in the opposite direction. The different ways to contemplate the limits on the productive capacity of a node are also discussed, together with the alternatives to model them and their overall effect on the dynamics of the supply chain. Other usual assumptions affecting the nature of the relationship with the customer, such as the impossibility to assume negative returns, or the case of lost sales (as opposed to the usual backlog assumption) are discussed. In order to provide the reader with additional insights, the results of the simulation of the mathematical models are provided. The chapter concludes with a summary of the topics discussed and with a list of annotated references related to the topics discussed in the chapter.

More specifically, in this chapter, we:

- Revisit some hypotheses in the basic SC related to the nature of the relationship of the customer, i.e. the case of lost sales as opposed to the backlog, and the impossibility by the customer to order negative returns (Sect. 6.2).
- Model and analyse the effect of variable lead times in SC dynamics (Sect. 6.3) and an important side-effect such as order crossover (Sect. 6.4).
- Study the effects of finite capacity (Sect. 6.5).

J. M. Framinan, *Modelling Supply Chain Dynamics*,
https://doi.org/10.1007/978-3-030-79189-6_6

6.2 Revisiting Some Hypotheses in the Basic SC Model

The small simulation model presented in Sect. 4.5.5 provides a tool to assess (even if in a limited manner) the effect of some simplifying assumptions done in the basic (analytical) model. Recall that the most critical assumptions are:

- The hypothesis of backlogged demand.
- The hypothesis that it is possible to issue negative orders (which, in this case, are interpreted as if the excess inventory is returnable to the supplier).

6.2.1 *Backlogged Demand*

An assumption adopted when developed the basic model in Sect. 4.5.1 is that, if there is no enough inventory, the final customer waits for the demand until it is fulfilled. This is equivalent to assuming that the inventory can be negative, since precisely the magnitude of the backlogged demand is represented by the amount of negative inventory. If this is not the case and the demand that cannot be met is lost (e.g. the customer finds the demand somewhere else, then the basic code in Sect. 4.5.5 can be altered to incorporate the non-linearity due to Eq. (4.5) in the following manner:

No backlog

```
...
# customer demand is statisfied
I[t] = max(I[t] - curr_demand,0)
...
```

and the differences between the two options (backlog versus no backlog) can be compared for a specific problem setting. In case of the setting in the basic code, the results are shown in Figs. 6.1 and 6.2 for the inventory and the orders, respectively. In both figures, the first 200 time periods are omitted to remove the transient effect.

As it can be seen from Fig. 6.1, there are substantial differences in terms of inventory variability. This is a foreseeable result since, in the no backlog case, inventory can be negative, and therefore, the variability is higher. Indeed, the resulting $NSAmp$ is smaller. However, we can see in Fig. 6.2 that this does not translate into a smaller order variability. Indeed, the resulting BWE is higher for the no backlog case.

6.2.2 *Returned Orders*

A hypothesis adopted in Sect. 4.5.3 is that the order size can be negative if it is found that the base stock level is smaller than the current inventory position (on-hand inventory plus work in process). Recall that the original formula is given by

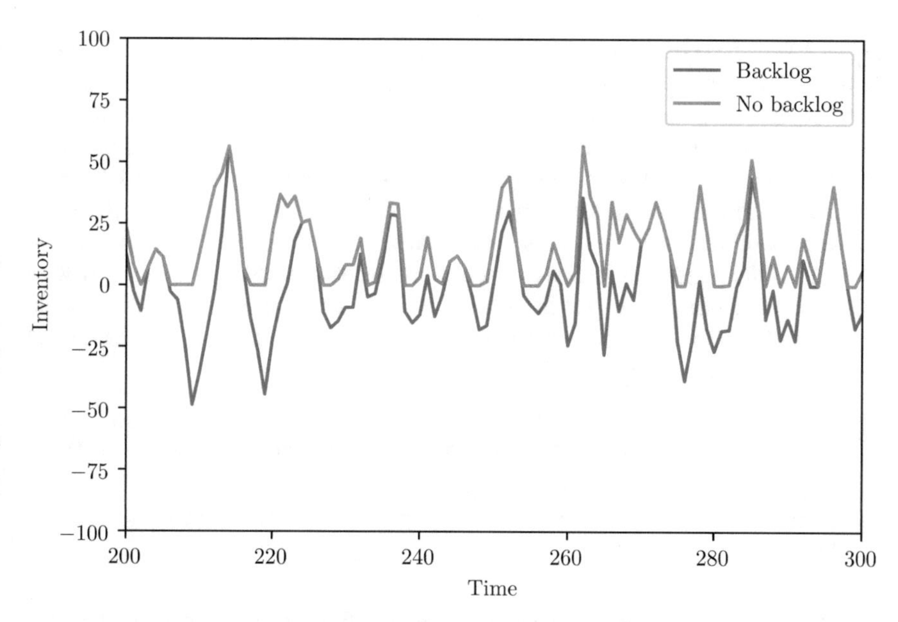

Fig. 6.1 Inventory variability under the hypotheses of backlog and no backlog

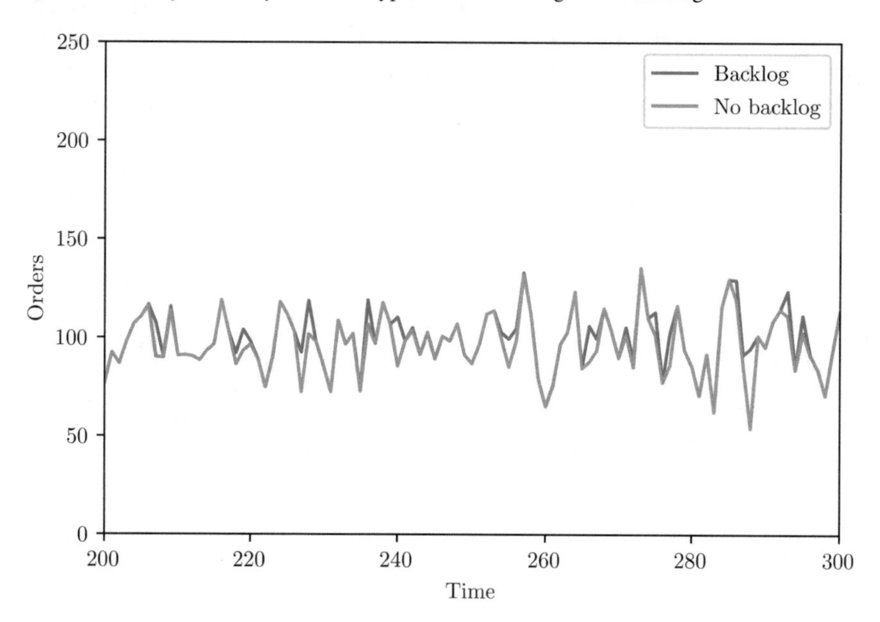

Fig. 6.2 Order variability under the hypotheses of backlog and no backlog

Eq. (4.12) and the hypothesis—also known as *reverse logistics*—allows that it can be linearised into Eq. (4.13) to make it tractable. Note that the hypothesis does not only requires that the supplier is willing to accept that the product can be returned, but also that it can be returned at no cost (otherwise this would influence the OUT policy, which would not be optimal). Clearly, the hypothesis does not seem to be quite realistic, and many authors using it had pointed out that there were no substantial differences in the results. Intuitively, one can see that, indeed the differences will be small if the number of periods with excess inventory is zero, and this is will be related (among other factors) to the variability of the demand and to the safety stock.

To check the effects of the assumption, the basic code in Sect. 4.5.5 can be altered to incorporate Eq. (4.12) in the following manner:

No returns

```
...
#calculates current order (moving average of m periods)
curr_order = max(moving_average(D,m,t)*L - I_prima[t] - W_prima[t],0)
...
```

Although it cannot be appreciated in Fig. 6.3, the BWE indicator obtained if returns are allowed almost doubles that if returns are not allowed. More modest differences are obtained in terms of $NSAmp$, as it can be seen in Fig. 6.4. The simulation parameters employed in Figs. 6.3 and 6.4 are summarised in Table 6.1. Note that, in both cases, the demand series is the same, in order to remove the possibility of different results due to the stochasticity of the models.

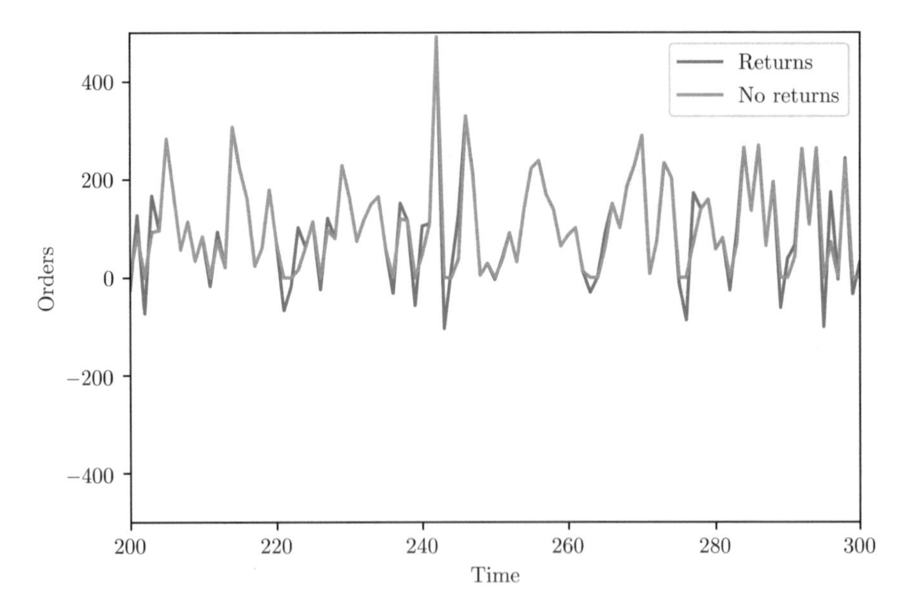

Fig. 6.3 Order variability under the hypotheses of returns and no returns

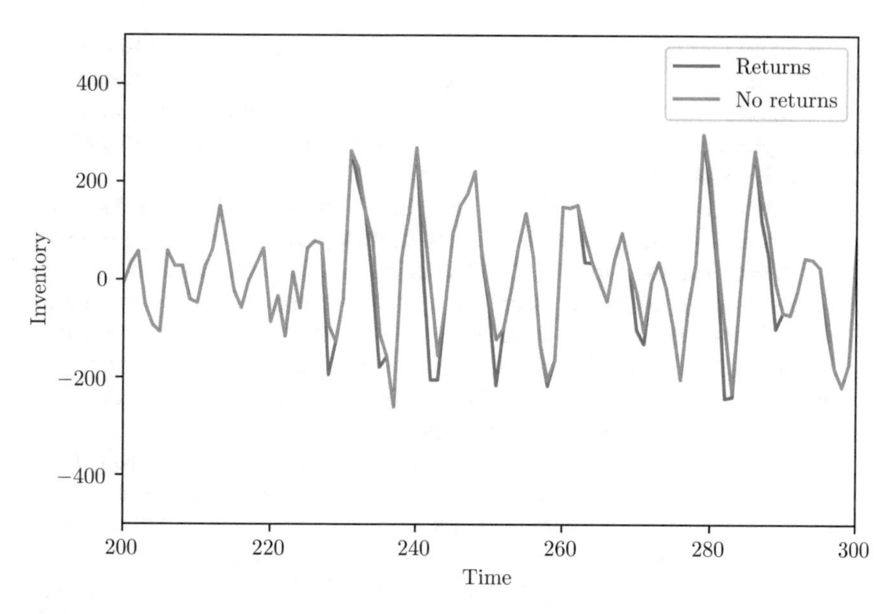

Fig. 6.4 Inventory variability under the hypotheses of returns and no returns

Table 6.1 Simulation parameters employed in the simulations in Fig. 6.3

Type of parameter	Parameter	Value
Simulation	Number of periods	20,000
	Initial inventory (I_0)	100
Supply chain	Safety factor	2
	Lead time	3
	Forecasting technique	MA($m = 3$)
	Demand distribution	Normal
	Average demand (μ)	100
	Standard deviation of the demand (σ)	50

6.3 Variable Lead Times

It has been observed that in many SC, the lead times are not deterministic. One of the causes cited is precisely the fact that some SCs are subject to capacity limits, an effect that would be discussed in Sect. 6.5. Other cited causes include the use of different transportation modes at different times (each one with different lead time characteristics), the use of alternative suppliers for the orders, and the use of expedited or emergency orders to increase the service level. Another case is the case where container lines in global SCs may take different routes, overtake each other at sea, stop in different ports along the way, or be held up for customs inspections at national borders. Finally, note that the current trend among companies to focus on customer service rather on cost minimisation is a major factor increasing the variability of the lead times, since, in order to fulfil the elevated customer service expectations, many companies have embraced a number of changes in order fulfilment strategies, including placing a greater number of small-sized orders and the use of different transportation modes to reach to their customers.

Given the effect of (constant) lead times in the bullwhip effect detected in the models in Chap. 4, it is foreseeable that lead times variability may have a profound effect on the dynamics of the SC. Recall that, in the OUT policies usually assumed in our models, ordering decisions are made so the base stock can face the demand occurring during the lead time, therefore, variable lead times add another source of uncertainty in order to calculate the base stock. In this section, we develop relatively simple analytical models to discuss this effect and then we give some hints on how the variable lead times can be represented in simulation models.

The main change with respect to the hypotheses in the model developed in Chap. 4 is that the lead times are not deterministic, so they can be considered as random variables. Therefore, L_t are now RVs describing the lead time in time period t. Furthermore, L_t is assumed to be stationary and iid distributed, with $E[L_t] = \mu_L$ and $V[L_t] = \sigma_L^2$. As usual, we assume that, as in the basic model, D_t follows a stationary first-order autoregressive process AR(1), i.e. $D_t = d + \rho \cdot D_{t-1} + \epsilon_t$. We also assume that, as in previous models, the node uses a base stock policy for inventory replenishment and can return to the provider the excess inventory without additional costs, and that the demand that cannot be met with the inventory is backlogged. In this case, we know from Eq. (4.15) in Chap. 4 that O_t the replenishment order to be issued at time period t is given by

$$O_t = s_t - s_{t-1} + D_t \tag{6.1}$$

where s_t is the base stock for period t given by the known expression in Eq. (4.11): $s_t = \hat{d}_t^L + z \cdot \hat{\sigma}_t^L$, i.e. the base stock is designed to cover the expected demand across the risk period ($\hat{d}_t^L$) plus its estimated standard deviation $\hat{\sigma}_t^L$ multiplied by the safety factor z.

In the case of variable lead times, L_t is no longer a known constant, but it also has to be estimated by the node to estimate the length of the risk period. For now, let us assume that the node is able to perform an MSSE estimate of L_t. In this case, since L_t are iid, we know that the MMSE estimate of L_t is simply its mean, i.e. μ_L.

Once that the length of the risk period has been estimated, the node has to estimate the demand across such risk period. If again we assume that the node is able to perform an MMSE estimate of the demand, we know from Eq. (4.20) in Chap. 4 that $\hat{d}_t^L$ the MMSE estimation of the demand across L time periods is given by

$$\hat{d}^L(t) = L\frac{d}{1-\rho} + \rho\frac{1-\rho^L}{1-\rho}\left(D_t - \frac{d}{1-\rho}\right) \tag{6.2}$$

so, in our case, we have

$$\hat{d}^L(t) = \mu_L \frac{d}{1-\rho} + \rho \frac{1-\rho^{\mu_L}}{1-\rho}\left(D_t - \frac{d}{1-\rho}\right) \tag{6.3}$$

Furthermore, under the previous hypotheses, we know that $\hat{\sigma}_t^L$ does not depend on t, so substituting in the expression for the replenishment orders, we have

$$O_t = \rho \frac{1-\rho^{\mu_L}}{1-\rho}(D_t - D_{t-1}) + D_t \tag{6.4}$$

or

$$O_t = \frac{1-\rho^{\mu_L+1}}{1-\rho} \cdot D_t - \frac{\rho-\rho^{\mu_L+1}}{1-\rho} \cdot D_{t-1} \tag{6.5}$$

From the above equation, indicator BWE can be obtained. Taking the variance of O_t, we have

$$V[O_t] = \left(\frac{1-\rho^{\mu_L+1}}{1-\rho}\right)^2 V[D] + \left(\frac{\rho-\rho^{\mu_L+1}}{1-\rho}\right)^2 V[D] - \tag{6.6}$$

$$2\frac{1-\rho^{\mu_L+1}}{1-\rho}\frac{\rho-\rho^{\mu_L+1}}{1-\rho} cov(D_t, D_{t-1}) \tag{6.7}$$

Since we know that $cov(D_t, D_{t-1}) = \rho \cdot V[D]$ as the demand is AR(1), BWE can be expressed as

$$BWE = \frac{V[O_t]}{V[D_t]} = \tag{6.8}$$

$$\frac{1}{1-\rho^2}\left((1-\rho^{\mu_L+1})^2 + (\rho-\rho^{\mu_L+1})^2 - 2(1-\rho^{\mu_L+1})(\rho-\rho^{\mu_L+1})\rho\right) \tag{6.9}$$

It can be seen that there is demand amplification if $\rho > 0$. The intensity of the amplification is related to μ_L, i.e. the highest the average lead times, the highest the bullwhip effect.

The case $\rho = 0$ (iid demands) reduces to the no amplification case where the demand is 'chased' ($O_t = D_t$). Note, however, that there is no dampening of the bullwhip effect as it happens in the corresponding case where capacity is understood as the rejection of orders higher than a given threshold.

As it can be seen, the formulae become more voluminous, and therefore, less instructive. Therefore, we only derive an additional model if both the demand and the lead times are estimated using the moving average in the case that both RVs are iid.

6.3.1 Estimating Both Demand and Lead Times Using the Moving Average

In this case, the estimate of the demand in the risk period is

$$\hat{D}_t^L = \left(\frac{1}{m_L}\sum_{j=1}^{m_L} L_{t-j}\right)\cdot\left(\frac{1}{m}\sum_{i=0}^{m-1} D_{t-i}\right) \tag{6.10}$$

where m_L and m are the number of periods that are used to estimate both the lead time and the demand according to the moving average. In this case, the order would be

$$O_t = D_t + \frac{1}{m \cdot m_L} \cdot \tag{6.11}$$

$$\left(\sum_{i=0}^{m-1} D_{t-i}L_{t-1} + (D_t - D_{t-m})\sum_{j=2}^{m_L} L_{t-j} - (\sum_{i=0}^{m-1} D_{t-i-1})L_{t-m_L-1}\right) \tag{6.12}$$

Here we can use the expression of the total variance to obtain $V[O_t]$, i.e.

$$V[O_t] = E[V[O_t|D_t]] + V[E[O_t|D_t]]$$

To obtain the first term of the equation, we first take the conditional variance of O_t

$$V[O_t|D_t] = \frac{V[L]}{m^2 \cdot m_L^2} \cdot \tag{6.13}$$

$$\left(\left(\sum_{i=0}^{m-1} D_{t-i}\right)^2 + (D_t - D_{t-m})^2(m_L - 2) + \left(\sum_{i=0}^{m-1} D_{t-i-1}\right)^2\right) \tag{6.14}$$

The expectation of the conditional variance is

$$E[V[O_t|D_t]] = \frac{V[L]}{m^2 \cdot m_L^2} \cdot \tag{6.15}$$

$$\left(m\mu V[D] + (m\mu)^2 + 2\,V[D](m_L - 2) + mV[D] + (m\mu)^2\right) \tag{6.16}$$

or

$$E[V[O_t|D_t]] = \frac{2\,V[L]}{m^2 \cdot m_L^2}\left(m^2\mu^2 + (m + m_L - 2)V[D]\right) \tag{6.17}$$

Taking the conditional expectation

$$E[O_t|D_t] = D_t + \frac{\mu_L}{m \cdot m_L}\left(\sum_{i=0}^{m-1} D_{t-i}\right) + \tag{6.18}$$

$$\frac{\mu_L(m_L-2)}{m \cdot m_L}(D_t - D_{t-m}) - \frac{\mu_L}{m \cdot m_L}\left(\sum_{i=0}^{m-1} D_{t-i-1}\right) \tag{6.19}$$

or

$$E[O_t|D_t] = \left(1 + \frac{\mu_L(m_L-1)}{m \cdot m_L}\right) D_t - \frac{\mu_L(m_L-1)}{m \cdot m_L} D_{t-m} \tag{6.20}$$

And the variance of the conditional expectation is

$$V[E[O_t|D_t]] = \left(1 + \frac{\mu_L(m_L-1)}{m \cdot m_L}\right)^2 V[D] + \left(\frac{\mu_L(m_L-1)}{m \cdot m_L}\right)^2 V[D] \tag{6.21}$$

Plugging all the results, we have

$$BWE = 1 + 2\frac{\mu_L(m_L-1)}{m \cdot m_L}\left(1 + \frac{\mu_L(m_L-1)}{m \cdot m_L}\right) + \tag{6.22}$$

$$(m + m_L - 2)\frac{2\sigma_L^2}{m^2 \cdot m_L^2} + \frac{2\sigma_L^2\mu^2}{\sigma^2 \cdot m_L^2} \tag{6.23}$$

Note that all components are positive and contribute to the bullwhip effect. The first factor amplifying the bullwhip effect is due to the MA forecast of the demand, although in this case, there is a factor $\frac{m_L-1}{m_L}$ affecting μ_L. The last term of the expression is due to the MA forecast of lead times. In addition, another factor increasing the variability of the orders is due to the crossed influence of the forecast. Note that this term is zero if a single period is used to forecast demand and lead times. However, this comes at the expense of increasing the other terms of the bullwhip effect.

6.3.2 Simulation Models

The two models developed in the previous section maintain all the hypotheses than the basic model in Chap. 4 with exception of the constant lead times (most notably, the possibility of backlog and returns discussed in Sect. 6.2). In order to remove them, it is necessary to develop simulation models, which can easily accommodate the variable lead times. Here the key issue is to appropriately model the lead times, as they should be integer variables. In most literature, they are modelled as (continuous) RV and then rounded to the integer. Most of the literature assumes a gamma distribution in the lead times, although in some other cases a truncated normal distribution is employed. The reason for the prevalence of the gamma distribution to model lead times is because

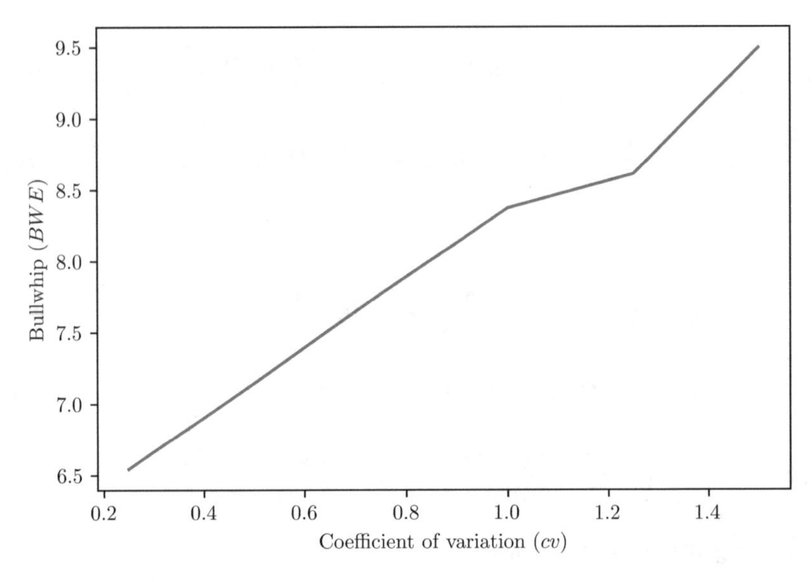

Fig. 6.5 Bullwhip effect increasing with lead time variability

this distribution only takes positive values (see Sect. B.4 in Appendix B). In addition, its two parameters can be linked to the mean and the variance, respectively, therefore, by using in the simulations a mean lead time and a coefficient of variation cv, it is possible to assess the effect of the variability of the lead times in the bullwhip effect, while the rest of the factors (including the average lead time) are maintained constant. In Fig. 6.5, this effect is depicted.

6.4 Order Crossover

A phenomenon that is intrinsically linked to the variability of the lead times discussed in Sect. 6.3 is the so-called *order crossover*. Order crossover refers to the case where the replenishments are received in a different sequence than they were ordered. This is a side-effect of lead time variability, so the product corresponding to a replenishment order issued after another *overtakes* another in-transit product corresponding to an order issued earlier and arrives before to the node's facility. Clearly, a consequence of order crossover is the fact that it distorts the incoming stream of replenishments.

Note that, even with stochastic lead times, order crossover does not need to happen if each replenishment is targeted for a specific order so the inventory is not indistinguishable. In other words, replenishment order placed in ith position[1] arrives in

[1] Here we change the most appropriate word 'order' by 'position' in order to avoid confusion between order as ordinality and order as a replenishment request issued by the retailer to the supplier.

ith order. This may happen because the variability of the lead times is small as compared to its mean and the physical situation of one order overtaking another does not happen. Alternatively, even the overtaking is produced during the transportation of the product, the orders are processed sequentially in the node's facilities, following a sort of First-In-First-Out (FIFO) rule.

If this is not the case, the Effective Lead Time or ELT (measured as the time between the ith order placed and the ith replenishment, or order arrival) may be different than the lead time of each order (measure as the time between when an order is issued and the time where this order arrives). Note that, in this case, i refers only to the ordinality, so the ith order placed does not have to arrive in the ith position of the replenishments. Figure 6.6 illustrate the concept.

It can be shown that, if ordering decisions are made every period, then the average lead time and the average expected lead time are the same, i.e. $E[L] = E[ELT]$, but the variance of the effective lead time is given by the following formula:

$$V[ELT] = E[L] - E[L^{1,2}] \tag{6.24}$$

where $L^{1,2}$ is the minimum between two random samples of L. The expression in Eq. (6.24) is dependent on the specific distribution of the RV L. Furthermore, for the case of the 1-period ordering decisions, it can be shown that $V[ELT]$ is bounded

$$E[L] - E[L^{1,2}] \leq V[ELT] \leq V[L] \tag{6.25}$$

The result with respect to the average ELT and its variance is more or less intuitive in Fig. 6.6, where it can be seen that the ELT are the same in the three periods considered (an average of 6 periods is obtained), whereas the actual lead

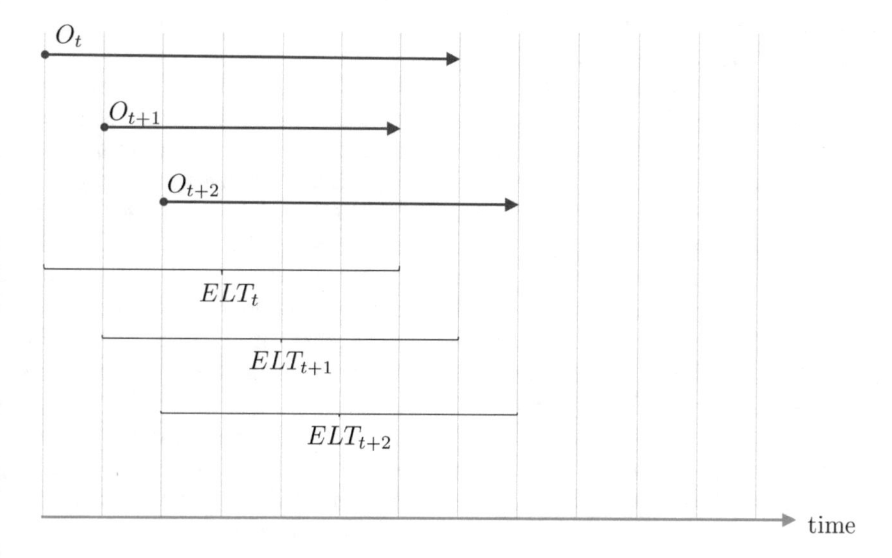

Fig. 6.6 Depiction of the concepts of lead time and ELT

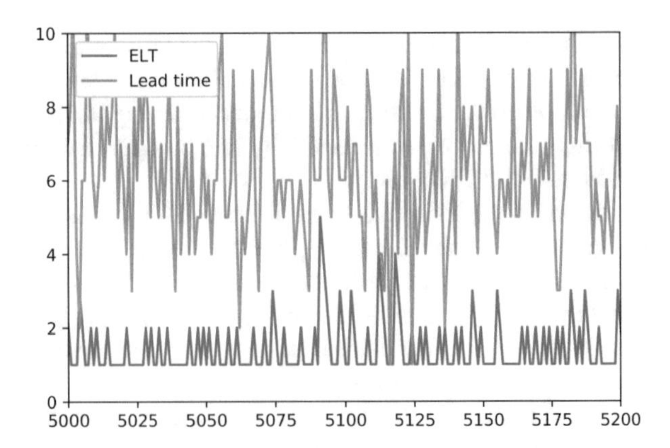

Fig. 6.7 Simulating lead time and ELT

Table 6.2 Simulation parameters employed in the simulations in Fig. 6.7

Type of parameter	Parameter	Value
Simulation	Number of periods	20,000
	Initial inventory (I_0)	100
Supply chain	Lead time distribution	Normal
	Average Lead time ($E[L]$)	6
	Lead time Standard deviation ($\sqrt{V[L]}$)	2
	Average demand (μ)	100
	Standard deviation of the demand (σ)	5

times are 7, 5 and 6, respectively. While the average is the same, the variability is naturally lower (zero in this case).

Order crossover is, in general, hard to be addressed with analytical models, or to be precise, it is hard to ignore order crossover. In fact, the model presented in this section do explicitly consider order crossover, since $R_t = O_{t+L_t}$ each replenishment is accounted for—and employed—whenever they arrive at the inventory, thus giving rise naturally to the order crossover phenomenon. However, it is not easy to compute the ELT, although it can be done using discrete-event simulation: in this case, the replenishment order of each period can be stored and, once the simulation has terminated, the ELT for each period can be computed by measuring the difference in time periods between the placement of ith order and the arrival of the ith replenishment. In Fig. 6.7, the differences between these two magnitudes are shown. The parameters employed for the model are given in Table 6.2.

One important consequence of order crossover is the fact that the replenishment policy adopted by the node must take into account that the effective lead time is smaller than the lead time (either the actual or the estimated lead time) and therefore the *effective risk period* is smaller. If an OUT policy is employed, we know then that s_t must be modified to take into account this fact, as $s_t = \hat{d}_t^L + z \cdot \hat{\sigma}_t^L$. Several alternatives (collectively denoted as *crossover-aware* OUT policies) can be used in this case

- Using $E\hat{L}T_t$ an estimate at time period t of the effective lead time to compute the risk period, so the base stock is now

$$s_t = \hat{d}_t^{E\hat{L}T_t} + z \cdot \hat{\sigma}_t^{E\hat{L}T_t} \tag{6.26}$$

  The estimation of ELT can be done similarly to the estimation of the demand, i.e. using any forecasting method such as the moving average, or the exponential smoothing.
- Since, according to Eq. (6.25), ELT can be bounded, its upper (lower) bound can be used to compute the risk period in order to provide an estimation of the worst (best) risk period that can be expected. Note that the computation of the bounds requires some assumption on the distribution of the lead time.
- To estimate the cdf of the demand during the effective risk period and use it to determine s_t by direct lookup instead of the normal assumption. Given the difficulty of obtaining an accurate estimate of this RV, this option represents a 'best case' and it has been used in the literature basically to assess the potential benefits of having crossover-aware replenishment policies.

It has been shown in different papers that the crossover-aware policies give better results in terms of order and inventory variability, even if the variability of the lead times is small, and therefore, order crossover seldom occurs. Figure 6.8 shows some sample results obtained with simulations using the parameters in Table 6.3.

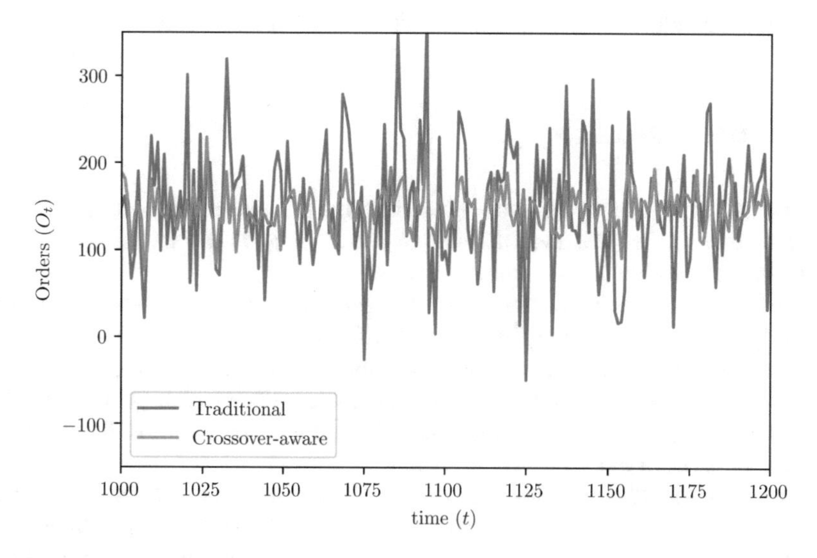

Fig. 6.8 Order variability in crossover-aware versus traditional OUT policies

Table 6.3 Simulation parameters employed in the simulations in Fig. 6.8

Type of parameter	Parameter	Value
Simulation	Number of periods	2,000
	Initial inventory (I_0)	100
Supply chain	Lead time distribution	Normal
	Average Lead time ($E[L]$)	6
	Lead time Standard deviation ($\sqrt{V[L]}$)	2
	Average demand (μ)	150
	Standard deviation of the demand (σ)	20
	Lead time estimation in Traditional	Moving average ($m = 10$)
	ELT estimation in Crossover-aware	Historical ELT data

With respect to the dynamic performance of crossover-aware policies, it has been shown that the most accurate the estimation of the effective risk period, the better the performance both in terms of order variability and inventory variability, i.e. the estimate of the cdf function (if feasible) yields better results than using all available historical data to forecast ELT, which, in turn, yields better results than using less data-intensive forecasting techniques such as the moving average. The superior performance of the estimation of the cdf function has been argued to be caused by the fact that the effective risk period does not generally follow a normal distribution, and therefore, the assumption that $s_t = \hat{d}_t^L + z \cdot \hat{\sigma}_t^L$ tends to underestimate the base stock required, particularly, if the variability of the lead times is high.

6.5 Capacitated SCs

Up to now, we have assumed that each node in the SC is operating with a so-called *infinite capacity*, i.e. whatever the size of the order is, it can be fulfilled by the upstream node within the (constant) lead time.

The main argument for assuming infinite capacity of the node is that the congestion effects in the supply chain may be negligible if the individual orders are as relatively small as compared to the capacity of the supply. Such supply chains are also called *exogenous* supply chains, as lead times are not influenced by variables internal to the system. Although this assumption may be realistic for some scenarios, it may not be so-well founded for others.

Modelling the dynamic effects of capacity limits in the SC may be quite complex since the *granularity* of both effects (bullwhip and capacity) are substantially different. In principle, capacity limits are the result of having a restricted number of resources (personnel, material, machinery, transportation, etc.) to carry out the operations required to complete the individual products within an order. Therefore, the natural unit to measure capacity is the product. In contrast, SC dynamics is usually observed at a more aggregated level, since we have usually measured the variations in the number of orders, each one composed, in general, of several products. The time scale of both phenomena is also quite different: the time periods considered for lead time and order replenishment are usually several orders of magnitude higher than the time required by one resource to complete an operation (typically days versus hours or even minutes).

Consequently, in order to integrate both aspects, there are basically two options:

- To develop an aggregated indicator of the shop floor capacity limits, so the capacity of the node can be expressed in terms of the number of orders that the node is capable to produce in a given time period.
- To develop a very detailed model of the shop floor capacity so the completion times of the individual orders are obtained. From these data, an aggregated lead time can be obtained (possibly taking into account transportation limits).

In this book, we deal only with the first approach. Note that the level of detail required for the second approach is very high and encompasses formulating many hypotheses regarding the way in which the operations are carried out in the node, thus making the so-obtained conclusions hardly generalizable, as they are extremely dependent on these assumptions. However, the first approach adopted here has the challenge of finding a workable indicator of the capacity in a SC node. In this regard, note that capacity limitations can be understood in (at least) two different manners:

1. **Flexible capacity**, where the node can invest in excess capacity to meet the requested order volume, possibly by subcontracting capacity or outsourcing some of the orders. In terms of the dynamics of the SC, this situation is equivalent to assuming infinite capacity, although it is clearly different from the economic viewpoint, as it can be assumed that there are extra costs involved in assuring the excess of demand over capacity.
2. **Inflexible capacity**, where the capacity is fixed and it cannot be adjusted to the demand. In this case, two different situations can be considered:

    - The node has the possibility to reject orders exceeding its regular capacity. This means that the arriving orders in excess of their capacity would be rejected. As a result, it is expected that the accepted orders can be delivered within the regular lead time, therefore, the lead time can be considered as constant.
    - The node does not reject orders exceeding its regular capacity, but may negotiate a specific (longer) lead time than the regular one, which may vary depending on the actual shop floor status.

At it can be seen, the term *capacity* of a node in the supply chain can be used to at least represents the following situations:

- Capacity as a limitation of the order size. In this case, the order size at any time cannot exceed a fixed capacity C. This case may model a situation where the node receiving the order has the ability to reject orders exceeding its regular capacity. Alternatively, it may model a situation where the node's capacity has to be booked much in advance so there is no room to allow orders beyond a certain excess. In practice, such constraint means rejecting some customers. As we will see in Sect. 6.5.1, if this is the case, the bullwhip effect is smoothed as the resulting orders have lesser variability than in the uncapacitated case.

- Capacity as a factor influencing the lead times, but that cannot be directly linked to current/past orders and/or demand. This is the case where the node cannot determine the precise lead time of the orders, but it should be estimated. In this case, the lead time in time t can be considered a stochastic variable independent of the demands, that has to be estimated by the node. This assumption represents an intermediate scenario where the node acknowledges some impact of the capacity, but the supply is still far greater than the demand, so it cannot be directly linked to it.
- Capacity that can be directly linked to current/past orders and/or demand. In this case, the node can guess the precise lead time of the order. Indeed, there would be no need to forecast the lead times.

Indeed, the concept of the capacity of a node may be linked to different policies or constraints of the node that can be captured with different models.

6.5.1 Modelling Capacity Limits

In this case, there is a limit C in the number of products that the node can process in a period, so the provider of the node never receives orders above this limit. Note that the most common approach to incorporate this limit into the model is to place a limit on the order quantity as determined by the OUT policy (see Eq. (4.15)). As a consequence, O_t^c the resulting order in the capacitated case is

$$O_t^c = \min\{O_t^u, C\} = \min\{s_t - s_{t-1} + D_t, C\} \tag{6.27}$$

where O_t^u is the order size in period t if the OUT policy would be applied in a SC without capacity limits (uncapacitated order size). Intuitively, it is easy to see (and it can be formally proved) that the variance of O_t is always equal to or smaller than O_t^u. Therefore, the bullwhip effect in the capacitated case is equal to or smaller than the bullwhip in the case of the uncapacitated case: $BWE^c = \frac{V[O_t^c]}{V[D_t]} \leq \frac{V[O_t^u]}{V[D_t]} = BWE^u$.

In some special cases, it can be shown that the bullwhip effect is dampened by the capacity limits: if the demands are assumed to be iid and they can be estimated MMSE, then it follows that $s_t = s_{t-1}$, and therefore, the expression of the bullwhip effect in the capacitated SC can be written

$$BWE^c = \frac{V[O_t^c]}{V[D_t]} = \frac{V[\min\{D_t; C\}]}{V[D_t]} \tag{6.28}$$

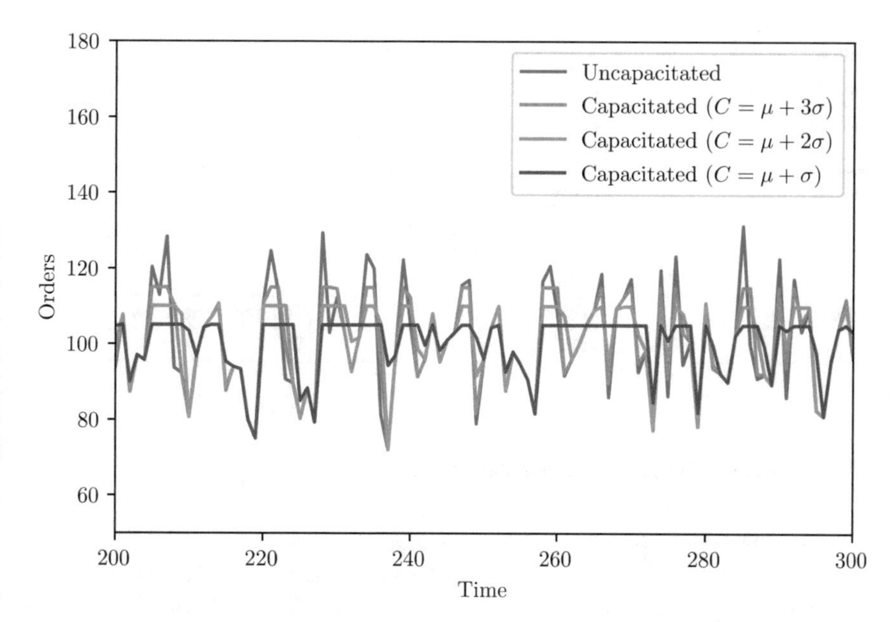

Fig. 6.9 Effect of capacity limits on orders variability

When the demand cannot be assumed to be iid and/or its estimation is not MMSE, the proofs can be more involved or even it can be hard to find a closed expression. However, note that Eq. (6.27) does not depend on these hypotheses, so the inequality $BWE^c \leq BWE^u$ still holds. Furthermore, the tighter the capacity constraints, the higher the difference between both cases. Figure 6.9 shows the results of a simulation of the variability of the orders for different values of C, expressed in terms of the average demand plus a number of times the standard deviation of the demand. The data employed for the simulation are given in Table 6.4

Again, it is easy to see that the variance of $\min\{D_t; C\}$ cannot be higher than that of D_t. As a result, $BWE \leq 1$ and it can be stated that, under these conditions, capacity (understood as the rejection of the orders in excess of the capacity) dampens the bullwhip effect (Fig. 6.9).

This positive behaviour with respect to BWE does not extend to the $NSAmp$ indicator. Indeed, it can be seen that $NSAmp$ increases with the capacity limits. Intuitively, this is caused by the additional backlogs induced in the periods that the demand exceeds the capacity limit, which causes a higher fluctuation in I_t: Recall that, due to the manner in which it has been modelled, I_t registers both the positive (*true* inventory) as well as the negative inventory (backlog). This speaks for the fact that this reduction in orders variability is achieved at the cost of decreasing service level and customer satisfaction.

Finally, it is to note that, in the previous model, the node is not operating under a modified OUT policy, which is not necessarily optimal in the capacitated case.

Table 6.4 Simulation parameters employed in the simulations in Fig. 6.9

Type of parameter	Parameter	Value
Simulation	Number of periods	5,000
	Initial inventory (I_0)	100
Supply chain	Lead time (L)	3
	Average demand (μ)	100
	Standard deviation of the demand (σ)	5
Policies	Estimation	Moving average ($m = 3$)
	Ordering policy	OUT

6.5.2 *Workload-Dependent Lead Times*

In some cases, it is not possible for the node to reject orders in excess of the capacity of the node. In this case, it is expected that the lead times of the demand are affected by the conditions in the shop floor. More specifically, it is known that there is a causal relationship between order release, capacity saturation and lead time. Theoretically, this relationship is expressed via Little's law and it has been well-documented in the observation of real-life manufacturing and transportation systems. Furthermore, specifically in the SC domain, empirical studies show the relationship between the lead times and the orders.

As a result, in general, the lead times of an order issued in period t are determined by the actual shop floor status. In general, we can think that the lead times of the orders to be released into the shop floor is a function of the pending work in the shop floor, i.e. $L_t = f(W_t)$, where f is node-specific, i.e. it models the ability of the particular node in the SC to absorb or not higher workloads without substantially modifying their lead times.

Also, note that this function may or may not be known to the decision makers in the node at the time to make order replenishment decisions. In general, we are assuming an OUT policy where the order-up-to level is the expected demand along the risk period (the expected lead time). If function $f(W_t)$ is known by the node, along with the corresponding work in process level in period t, then the risk period is known. Otherwise, the risk period must be estimated, possibly based on past data. Therefore, in general, together with L_t, we must also consider $\hat{L}_t$ the estimated lead time for orders that are to be issued in period t.

As a consequence, the sequence of events in the basic model described in Sect. 4.5 has to be modified accordingly. More specifically, at period t, the following sequence of events take place:

1. The retailer receives from the supplier the orders, i.e. R_t the replenishment taking place in period t arrives at the node. Note, however, that, in this case, $R_t \neq O_{t-L}$ as in the basic model. Indeed, the time of the arrival of each order would be computed depending on the actual status of the shop floor. The replenishment serves to update the inventory and the work in process.

2. The retailer receives the final customer demand and tries to satisfy it with the inventory at hand, thus decreasing the inventory. This step is the same than in the basic model.
3. The retailer studies her needs to face the future customer demand during the actual (real or estimated) lead time in view of the shop floor status, and consequently, she places an order to the supplier. This order will arrive after the actual (real) lead time, i.e. $R_{t+L_t} = O_t$. The order increases the work in process.

The set of equations modelling the first and second steps above sequence of actions are the following:

$$I_t = I_{t-1} + R_t - D_t \tag{6.29}$$

$$W_t = W_{t-1} - R_t \tag{6.30}$$

With respect to the replenishment decisions, if the function $L_t = f(W_t)$ is known to the Decision Maker, then he/she must simply forecast the demand over a known risk period. If, for instance, the demand can be considered iid and an MMSE estimation of the demand is carried out, then the order issued at period t is

$$O_t = \mu \cdot L_t - I_t - W_t \tag{6.31}$$

Otherwise, $\hat{L}_t$ an estimation of the length of the risk period must be obtained so the order issued would be

$$O_t = \mu \cdot \hat{L}_t - I_t - W_t \tag{6.32}$$

Once the order is issued, it increases the work in process, as in the basic SC model

$$W_t = W_t + O_t \tag{6.33}$$

Finally, a period for the reception of the order issued at period t has to be set, i.e.

$$R_{t+L_t} = R_{t+L_t} + O_t \tag{6.34}$$

Note in Eq. (6.34) that, depending on how $f(W_t)$ is modelled, it might be that orders issued at different periods arrive in the same period, therefore, the need of summing all the orders for the period. This did not happen in the models assuming constant lead times as each order arrives in a different period.

As it can be seen, the above set of equations allows to model the influence of capacity on lead times once $L_t = f(W_t)$ and a method for the Decision Maker to estimate L_t in case that the function is not known to him/her are defined. These issues are discussed in the next sections.

6.5.2.1 Models for L_t

One option is to model the lead time in one period as a function of the work in process using the so-called cycle time-throughput (CT-TP) curves. CT-TP are empirical curves used in practice to capture the relationship between the average cycle time (time for a job to be completed in a process) and the throughput rate of the process. Usually, these curves exhibit a *hockey stick* shape indicating the lead times are approximately constant if the intended throughput of a system is below a certain level, but it increases sharply beyond that level. Figure shows the typical shape of a CT-TP curve.

In view of this empirical behaviour of the CT-TP curve, we can develop a function mimicking this behaviour. One function proposed in the literature is

$$L_t = \begin{cases} L_0 & W_t \leq \varphi \\ L_0 \cdot e^{-\frac{1}{\eta}\left(1-\frac{W_t}{\varphi}\right)} & W_t > \varphi \end{cases} \tag{6.35}$$

In Eq. (6.35), when the work in process is below a threshold limit φ (denoted as saturation limit), the orders can be fulfilled at a constant lead time L_0 (base lead times). When W_t is above φ, the actual lead times increase in an exponential manner depending on two parameters η and φ. Parameter φ (saturation limit) is linked to the maximum number of work in process that the system can process without experiencing an increase in their lead times, i.e. it is related to the capacity of the node. On the other hand, η is a parameter that can be linked to the volume responsiveness of the node, i.e. its ability to cope with an increase of work in process beyond the saturation limit. If $\eta = 0$, then the node is not able to cope with a work in process beyond the saturation limit, and therefore, it collapses. Higher values of η indicate a better capability of the system to handle higher work in process. As it can be seen in Fig. 6.10, the trend of the resulting curve is similar to that of the CT-TP curve.

Finally, it has to be noted that the lead times are integer numbers (periods), therefore, they have to be rounded to the closest integer so they can be plugged into Eq. (6.34).

6.5.2.2 Methods to Estimate $\hat{L}_t$

Since the CT-TP curves can be obtained in an empirical manner from the observation of the shop floor data, we can assume that the node may have this information at hand so the risk period can be accurately estimated. If this is the case, the *true* value of L_t can be used to obtain the replenishment order, e.g. under an OUT policy, we would have

$$O_t^c = \hat{d}_t^{L_t} - I_t - W_t \tag{6.36}$$

where $\hat{d}_t$ is the estimation of the demand across the future L_t periods.

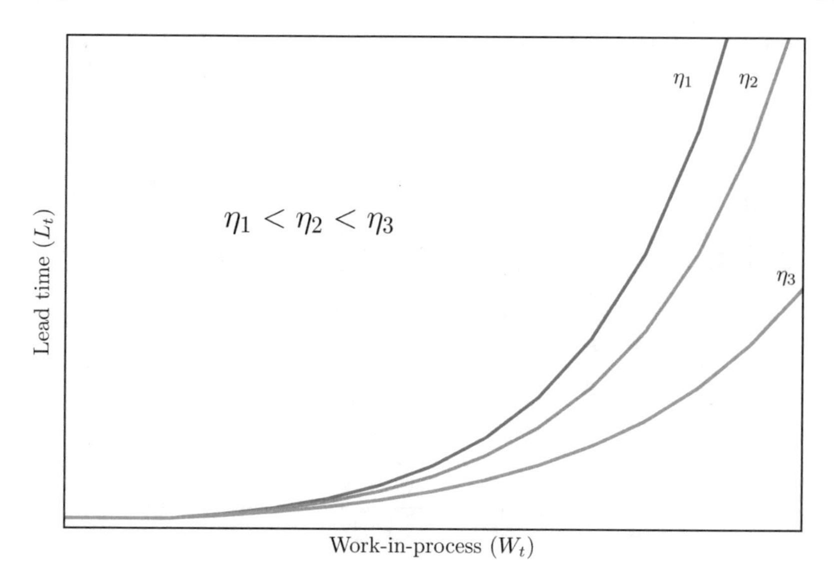

Fig. 6.10 Emulation of CT-TP curve using Eq. (6.35)

However, it might be that the CT-TP is not known to the decision maker, so he/she may have to estimate L_t, possibly based on past data. If this is the case, the available method for demand estimation (moving average, exponential smoothing, etc.) may be used to estimate it.

6.5.3 Combining Capacity Limitation with Load-Depending Lead Times

One problem of using Eq. (6.35) to model the lead times is that, in many real-life situations, there are limits to the orders that an overloaded system can accept. Therefore, if a shop floor has a high work in process, it is likely that it would reject new orders, as these would have lead times unacceptably higher for the customer. In this manner, we can combine the idea of rejecting orders above a certain capacity limit with the increasing lead times generated by a higher work in process. More specifically, we consider a node with a base (minimum) lead time L_0 and a maximum admissible lead time L_{max}. The node can process ν units per period, therefore, it is not possible for the node to process more than $\nu \cdot L_{max}$ units without exceeding the maximum lead time, so future orders would have to accommodate this value. This can be done in a two-step procedure: In the first step, O_t^u is computed as in an uncapacitated node. In a second step, the actual (capacitated) orders O_t^c are computed according to the following equation:

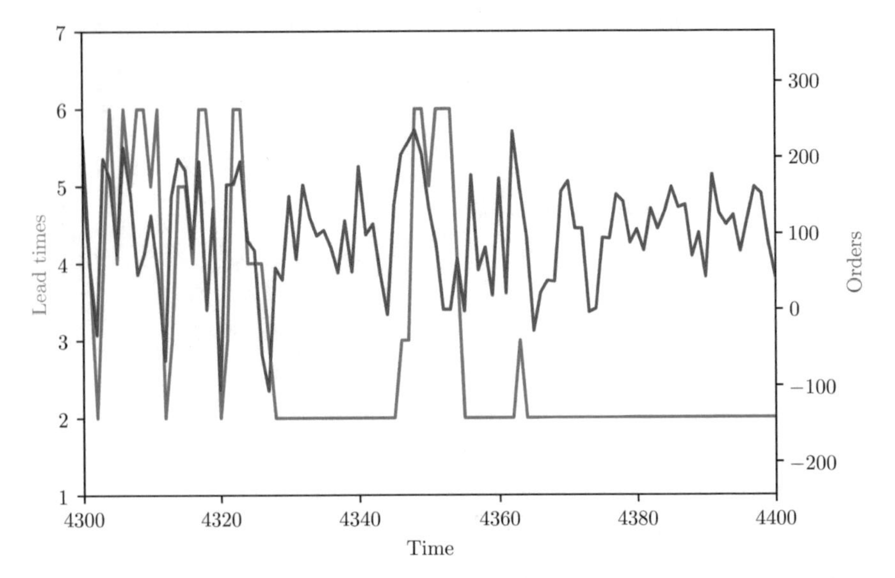

Fig. 6.11 Load-dependent lead times with maximum lead times: Lead-times versus Orders

$$O_t^c = \begin{cases} O_t^u & \text{if } O_t^u + W_t \leq \nu \cdot L_{max} \\ \max\{\nu \cdot L_{max} - W_t, 0\} & \text{otherwise} \end{cases} \tag{6.37}$$

By establishing these limits, the actual lead times of an order O_t^c issued at period t would be given by the following equation:

$$L_t = \frac{W_t + O_t}{\nu} \tag{6.38}$$

Note that, again, L_t values must be rounded to the ceiling integer in order to be plugged into Eq. (6.34). Combining these ideas, a stable representation of work-load dependent lead times can be achieved. In Figs. 6.11 and Fig. 6.12, the effect of the variability can be observed with the spikes in the lead times whenever the work in process increase, and how the orders are reduced whenever the maximum lead time has been attained.

Indeed, this model shows again the characteristics of the capacitated SC: the BWE indicator can be controlled using L_{max} (since the excess orders would be rejected), and therefore, it could dampen the bullwhip effect, but at the price of increasing the $NSAmp$ indicator. Ultimately, when $L_{max} = L_0$, then the results of the model closely resemble those obtained for the model obtained in Sect. 6.5.1.

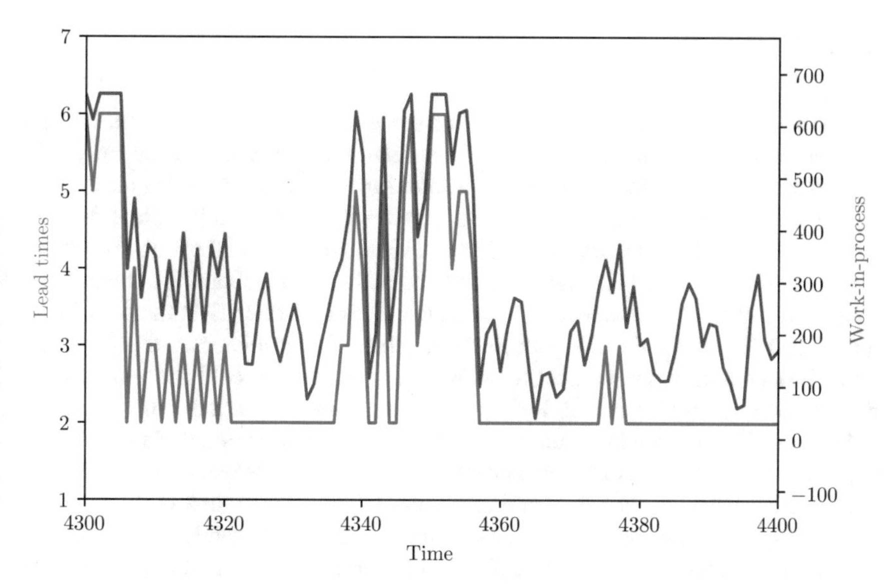

Fig. 6.12 Load-dependent lead times with maximum lead times: Lead-times versus Work in process

6.6 Conclusions

In this chapter, we have removed some of the common hypotheses assumed in the basic model developed in Chap. 4 and have also extended some limitations of this model. Regarding the two commonly adopted hypotheses (possibility of backlog and possibility of returns), it can be seen that the first one underestimates the bullwhip effect, whereas the effect of the second one is the opposite. In some sense, something similar happens with the variable lead times: while it is clear that variable lead times increase the bullwhip effect, the order crossover, a phenomenon that it is closely linked to this variability, offers opportunities for bullwhip reduction, particularly if crossover-aware replenishment policies are adopted.

Finally, the effect of capacity in SCs has been studied. Whenever a fixed capacity is assumed (i.e. the orders cannot exceed a certain threshold or capacity limit), the bullwhip effect is dampened, but the variability of the inventories increases. Therefore, there is a trade-off between both indicators and it should be some capacity limit where the operation of the node in the SC is optimal with respect to the costs.

6.7 Further Readings

The enrichment of the basic SC models and the study on their implications is being a constant in the SC literature. As we have seen, most of these enhancements of the initial models lead to non-linearities and/or further complexities that make them intractable in a closed form. Therefore, most of the literature on this topic is devoted to simulation models. Regarding the effect of returned orders in the bullwhip effect, the pioneering paper is [1], who show that the bullwhip effect can be notable over-estimated if the excess orders can be returned, a result further elaborated by [2] in the context of a non-serial SC. Note, however, that [3] have addressed this issue in a rather limited setting, showing that there were no great differences in considering the possibility of returns, or not. In Refs. [4, 5], the complex behaviour emerging from even simple SCs without the hypothesis of negative orders is explored. The difference in terms of amplification of orders and inventory variability between the assumption of lost sales instead of the (classical) backlogged demand is studied in [6].

An early paper on lead time variability analytically is [7], while one of the first simulation models to address it is [8]. Some causes for lead time variability are described in [9]. The impact of ignoring lead time variability in inventory replenishment policies is discussed in [10], while in [11] or in [12] the issue of whether it is better to reduce average lead time or the variance of the lead time is investigated in the SC context. Nielsen and Michna [13] show the suitability of modelling lead times as iid times after analysing the data from a real SC. The need of estimating average lead times to make replenishment decisions and its influence on the bullwhip effect is discussed in [14]. A measure of the bullwhip effect when the lead times are variable and the weighted moving average forecasting technique is employed is given in [15]. The specific case where the lead times follow an exponential distribution is analysed in [16] in terms of costs. Other papers analysing variable lead times in the SC context are [17–19], while [20] addresses the effect of variable lead times in SC risk management.

Regarding order crossover, earlier papers in the topic are [21, 22]. The proof of the mean and variance of the effective lead time (ELT) when the reordering period is 1 is given in [23] while the bounds on the variance of ELT are given in [24], also for one-period reordering decisions. A procedure to compute ELT using event-driven simulation similar to the one presented in Sect. 6.4 is presented in [25], where an in-depth analysis of the different crossover-aware strategies for OUT policies is discussed. In [9, 26], the use of the proportional controller as a strategy for mitigating inventory and order amplification in the presence of order crossover is discussed. Other papers exploring order crossover are [27–30].

With respect to the literature on capacitated SCs, arguments for the plausibility of assuming exogenous SCs can be found in, e.g. [31]. The consideration of capacity as a threshold can be read in [32–35]. The notions of flexible and inflexible capacity are taken from [36, 37]. Empirical evidence regarding the link between orders and supply lead times are discussed in [38]. Papers modelling SC with load-dependent lead times are [37, 39–41]. The use of CT-TP curves (see e.g. [42] for the concept) to model load-dependent processing times can be found in [43]. The use of the curves in Eq. (6.35) has been first proposed in [41]. Other studies dealing the capacity-constrained SCs are [44–46], and a recent contribution dealing with the detailed modelling of the productive capacity in SCs is [47].

Finally, other enhancements that have been studied in the literature address different aspects related to the products in the SC. These include the case of multiple-products SCs: in [48–50], the bullwhip effect in the case of multiple products with interdependent demand is addressed, whereas in [51], multivariate demand models are assumed). Furthermore, in [52], a multi-product, multi-stage SC is simulated. Other aspects related to the products discussed in the literature include the influence of the perishability of the products in the bullwhip effect [53], the use of substitute products [54], or the effect of the life-cycle demand [55].

References

1. Chatfield, D., Pritchard, A.: Returns and the bullwhip effect. Transp. Res. Part E: Logist. Transp. Rev. **49**(1), 159–175 (2013)
2. Dominguez, R., Cannella, S., Framinan, J.: On returns and network configuration in supply chain dynamics. Transp. Res. Part E: Logist. Transp. Rev. **73**, 152–167 (2015)
3. Chen, F., Ryan, J., Simchi-Levi, D.: The impact of exponential smoothing forecasts on the bullwhip effect. Naval Res. Log. **47**(4), 269–286 (2000)
4. Wang, X., Disney, S., Wang, J.: Stability analysis of constrained inventory systems with transportation delay. Eur. J. Oper. Res. **223**(1), 86–95 (2012)
5. Wang, X., Disney, S., Wang, J.: Exploring the oscillatory dynamics of a forbidden returns inventory system. Int. J. Product. Econom. **147**(PART A), 3–12 (2014)
6. Disney, S., Ponte, B., Wang, X.: Exploring the nonlinear dynamics of the lost-sales order-up-to policy. Int. J. Product. Res. (2020)
7. So, K., Zheng, X.: Impact of supplier's lead time and forecast demand updating on retailer's order quantity variability in a two-level supply chain. Int. J. Product. Econom. **86**(2), 169–179 (2003)
8. Chatfield, D., Kim, J., Harrison, T., Hayya, J.: The bullwhip effect—impact of stochastic lead time, information quality, and information sharing: a simulation study. Product. Operat. Manag. **13**(4), 340–353 (2004)
9. Wang, X., Disney, S.M.: Mitigating variance amplification under stochastic lead-time: the proportional control approach. Europ. J. Operat. Res. **256**(1), 151–162 (2017)
10. Bagchi, U., Hayya, J., Chu, C.H.: The effect of lead-time variability: the case of independent demand. J. Operat. Manag. **6**(2), 159–177 (1986)
11. Heydari, J., Baradaran Kazemzadeh, R., Chaharsooghi, S.: A study of lead time variation impact on supply chain performance. Int. J. Adv. Manuf. Technol. **40**(11–12), 1206–1215 (2009)
12. Chaharsooghi, S., Heydari, J.: Lt variance or lt mean reduction in supply chain management: which one has a higher impact on sc performance? Int. J. Product. Econom. **124**(2), 475–481 (2010)
13. Nielsen, P., Michna, Z.: The impact of stochastic lead times on the bullwhip effect—an empirical insight. Manag. Product. Eng. Rev. **9**(1), 65–70 (2018)
14. Michna, Z., Nielsen, P., Nielsen, I.: The impact of stochastic lead times on the bullwhip effect-a theoretical insight. Product. Manuf. Res. **6**(1), 190–200 (2018)
15. Nakade, K., Aniyama, Y.: Bullwhip effect of weighted moving average forecast under stochastic lead time, pp. 1277–1282 (2019)

16. He, X., Kim, J., Hayya, J.: The cost of lead-time variability: the case of the exponential distribution. Int. J. Product. Econom. **97**(2), 130–142 (2005)
17. Duc, T., Luong, H., Kim, Y.D.: A measure of the bullwhip effect in supply chains with stochastic lead time. Int. J. Adv. Manuf. Technol. **38**(11–12), 1201–1212 (2008)
18. Do, N., Nielsen, P., Michna, Z., Nielsen, I.: Quantifying the bullwhip effect of multi-echelon system with stochastic dependent lead time. IFIP Adv. Inf. Commun. Technol. **438**(PART 1), 419–426 (2014)
19. Ponte, B., Costas, J., Puche, J., Pino, R., de la Fuente, D.: The value of lead time reduction and stabilization: a comparison between traditional and collaborative supply chains. Transport. Res. Part E: Log. Transp. Rev. **111**, 165–185 (2018)
20. Bandaly, D., Satir, A., Shanker, L.: Impact of lead time variability in supply chain risk management. Int. J. Product. Econom. **180**, 88–100 (2016)
21. Zalkind, D.: Order-level inventory systems with independent stochastic leadtimes. Manag. Sci. **24**(13), 1384–1392 (1978)
22. Robinson, L., Bradley, J., Thomas, L.: Consequences of order crossover under order-up-to inventory policies. Manuf. Serv. Oper. Manag. **3**(3), 175–188 (2001)
23. Hayya, J.C., Bagchi, U., Kim, J.G., Sun, D.: On static stochastic order crossover. Int. J. Product. Econom. **114**(1), 404–413 (2008)
24. Hayya, J.C., Bagchi, U., Ramasesh, R.: Cost relationships in stochastic inventory systems. Int. J. Product. Econom. **130**(2), 196–202 (2011)
25. Chatfield, D.C., Pritchard, A.M.: Crossover aware base stock decisions for service-driven systems. Transp. Res. Part E: Logist. Transp. Rev. **114**, 312–330 (2018)
26. Disney, S., Maltz, A., Wang, X., Warburton, R.: Inventory management for stochastic lead times with order crossovers. Europ. J. Operat. Res. **248**(2), 473–486 (2016)
27. Bradley, J., Robinson, L.: Improved base-stock approximations for independent stochastic lead times with order crossover. Manuf. Serv. Operat. Manag. **7**(4), 319–329 (2005)
28. Robinson, L., Bradley, J.: Further improvements on base-stock approximations for independent stochastic lead times with order crossover. Manuf. Serv. Operat. Manag. **10**(2), 325–327 (2008)
29. Wensing, T., Kuhn, H.: Analysis of production and inventory systems when orders may cross over. Ann. Operat. Res. **231**(1), 265–281 (2015)
30. Yang, C.C., Lin, D.: Stochastic lead time with order crossover. Qual. Technol. Quant. Manag. **16**(5), 575–587 (2019)
31. Hum, S.H., Parlar, M.: Measurement and optimization of supply chain responsiveness. IIE Trans. **46**, 1–22 (2014)
32. Evans, G., Naim, M.: The dynamics of capacity constrained supply chains. Proceedings of International System Dynamics Conference, pp. 28–35 (1994)
33. Cannella, S., Ciancimino, E., Marquez, A.C.: Capacity constrained supply chains: a simulation study. Int. J. Simul. Process Modell. **4**(2), 139 (2008)
34. Chen, L., Lee, H.L.: Bullwhip effect measurement and its implications bullwhip effect measurement and its implications. Operations **60** (4)(March 2014), 771–784 (2012)
35. Ponte, B., Wang, X., Fuente, D.D., Disney, S.M.: Exploring nonlinear supply chains: the dynamics of capacity constraints. Int. J. Product. Res. **7543** (2017)
36. Boute, R.N., Disney, S.M., Lambrecht, M.R., Van Houdt, B.: A win-win solution for the bullwhip problem. Product. Plann. Control **19**(7), 702–711 (2008)
37. Boute, R.N., Disney, S.M., Lambrecht, M.R., Van Houdt, B.: Designing replenishment rules in a two-echelon supply chain with a flexible or an inflexible capacity strategy. Int. J. Product. Econom. **119**(1), 187–198 (2009)
38. Nielsen, P., Michna, Z.: An approach for designing order size dependent lead time models for use in inventory and supply chain management. In: Intelligent Decision Technologies 2016, pp. 15–25. Springer International Publishing, Cham (2016)
39. Helo, P.: Dynamic modelling of surge effect and capacity limitation in supply chains. Int. J. Product. Res. **38**(17), 4521–4533 (2000)
40. Framinan: Capacity considerations in the bullwhip effect in supply chains: the effect on lead times. In: Paper Presented at the 11th Conference on Stochastic Models of Manufacturing and Service Operations, June 4–9, 2017, Acaya, Italy (2017)

41. Cannella, S., Dominguez, R., Ponte, B., Framinan, J.: Capacity restrictions and supply chain performance: modelling and analysing load-dependent lead times. Int. J. Product. Econom. **204**, 264–277 (2018)
42. Ankenman, B., Bekki, J., Fowler, J., Mackulak, G., Nelson, B., Yang, F.: Simulation in production planning: an overview with emphasis on recent developments in cycle time estimation. Int. Ser. Operat. Res. Manag. Sci. **151**, 565–591 (2011)
43. Moench, L., Fowler, J., Mason, S.: Production planning and control for semiconductor wafer fabrication facilities: modeling, analysis, and systems. Modeling, Analysis, and Systems, Production Planning and Control for Semiconductor Wafer Fabrication Facilities (2013)
44. Buchmeister, B., Friscic, D., Palcic, I.: Bullwhip effect study in a constrained supply chain, pp. 63–71 (2014)
45. Lin, W.J., Jiang, Z.B., Wang, L.: Modelling and analysis of the bullwhip effect with customers baulking behaviours and production capacity constraint. Int. J. Product. Res. **52**(16), 4835–4852 (2014)
46. Hussain, M., Khan, M., Sabir, H.: Analysis of capacity constraints on the backlog bullwhip effect in the two-tier supply chain: a taguchi approach. Int. J. Logist. Res. Appl. **19**(1), 41–61 (2016)
47. Costa, A., Cannella, S., Corsini, R., Framinan, J., Fichera, S.: Exploring a two-product unreliable manufacturing system as a capacity constraint for a two-echelon supply chain dynamic problem. Int. J. Product. Res. (2020)
48. Sadeghi, A.: Providing a measure for bullwhip effect in a two-product supply chain with exponential smoothing forecasts. Int. J. Product. Econom. **169**, 44–54 (2015)
49. Raghunathan, S., Tang, C., Yue, X.: Analysis of the bullwhip effect in a multiproduct setting with interdependent demands. Operat. Res. **65**(2), 424–432 (2017)
50. Raghunathan, S., Tang, C., Yue, X.: Bullwhip effect of multiple products with interdependent product demands. Int. Ser. Oper. Res. Manag. Sci. **276**, 145–161 (2019)
51. Nagaraja, C., McElroy, T.: The multivariate bullwhip effect. Europ. J. Operat. Res. **267**(1), 96–106 (2018)
52. Wangphanich, P., Kara, S., Kayis, B.: Analysis of the bullwhip effect in multi-product, multi-stage supply chain systems-a simulation approach. Int. J. Product. Res. **48**(15), 4501–4517 (2010)
53. Minner, S., Transchel, S.: Order variability in perishable product supply chains. Europ. J. Operat. Res. **260**(1), 93–107 (2017)
54. Duan, Y., Yao, Y., Huo, J.: Bullwhip effect under substitute products. J. Operat. Manag. **36**, 75–89 (2015)
55. Nepal, B., Murat, A., Babu Chinnam, R.: The bullwhip effect in capacitated supply chains with consideration for product life-cycle aspects. Int. J. Product. Econom. **136**(2), 318–331 (2012)

Chapter 7
Closed-Loop Supply Chain

7.1 Introduction

A Closed-Loop Supply Chain (CLSC) is a supply chain that includes the recollection and recovery of their used products. This may include some of the following activities:

- Repair, so the recollected product is brought back to a functional condition by replacing or servicing their damaged components.
- Reuse, so the recollected product can be used (possible in a secondary market) without conducting repair activities.
- Recycling, so the recollected product (or some of their components) is processed so they may be used as materials to make a new product.
- Remanufacturing (better known as rebuilding in certain sectors), so the recollected product is restored to a state where it can be considered as new in the sense that it fulfils the same standards as a new product. As such, it may require the repair and/or the replacement of its non-functional components.

As depicted in figure, the recollection and recovery processes create an additional flow of material, information and money that goes upwards in the SC.

Among the recovery processes described above, remanufacturing has become one of the key ones, as it retains the whole value of the product (in contrast to repairing or reusing) by delivering a product whose performance is at least to its original specifications. Often cited products where the remanufactured part is indistinguishable from the original are the single-use cameras, leasing of office equipment or spare parts in the automotive and telecommunications industry.

There are a number of drivers fostering remanufacturing. Clearly, there are important environmental benefits in remanufacturing, and, since societies are moving towards more environmentally conscious, environmental regulations are being passed. Despite these advantages, there are also important challenges in remanufacturing. Some of them are of technical nature, as the remanufacturing process may be technically complex or may require important changes in the product design. However, in many cases, the challenges are of operational nature and are motivated

J. M. Framinan, *Modelling Supply Chain Dynamics*,
https://doi.org/10.1007/978-3-030-79189-6_7

basically by the complexity of adequately managing and integrating the direct and reverse flows of materials and information in order to make profitable the supply network. Precisely, as we know from previous chapters, one of the phenomena that most hurt the profitability of the supply chain is the bullwhip effect; therefore, it is of interest to study how this effect works in CLSC when remanufacturing processes have to be taken into account.

In this chapter, we:

- Discuss different remanufacturing scenarios, with a focus on the so-called hybrid manufacturing-remanufacturing system (Sect. 7.2).
- Present a simple, two-echelon model of the hybrid manufacturing-remanufacturing scenario (Sect. 7.2).
- Discuss how the information transparency and the visibility of the remanufacturing operations may influence the dynamic performance of the SC (Sect. 7.3).
- Present future research lines including additional factors such as the variability in the returns (Sect. 7.4).

7.2 Remanufacturing Scenarios

Remanufacturing may take different forms in practice. In this chapter, we focus on the so-called hybrid manufacturing-remanufacturing scenario, where integrated management of both types of operations is carried out. Although using third parties for remanufacturing may be an option in many industries, an increasing number of companies are getting involved in the remanufacturing of their products. Furthermore, in general, remanufacturing is less costly than the manufacture of a new product; therefore, it seems interesting to try to fulfil all the demands by remanufacturing and then use manufacture to cover the deficit. This strategy denoted as Priority-To-Remanufacture (PTR), as opposed to the Priority-To-Manufacture (PTM) strategy, would use remanufacturing as a secondary resource to cover the demand that cannot be met by the manufacturing process. Figure 7.1 shows the main elements of a node following this scenario: the demand from the customers is met in a make-to-stock fashion from an inventory of products. This inventory is constituted both by (undistinguishable) new and remanufactured products. Products sold to the customer (or to the next node in the supply chain) are used and a fraction of them (the so-called return rate denoted by α) are returned for remanufacturing. We denote by L_C (consumption lead time) the time that it takes for the products to be consumed. In the first basic model, we will assume that this consumption lead time is a constant, although the effect of its variability would be discussed in Sect. 7.4.

Once the product has been returned to the company, it is remanufactured. This process takes a time which is denoted by L_R (remanufacturing lead time), and, once it is finished, the remanufactured product is stored in the inventory so it is available for future sales. There are different studies citing the return rate in a range between 0.25 and 0.75, so it is very unlikely that (even if this would be in many cases the

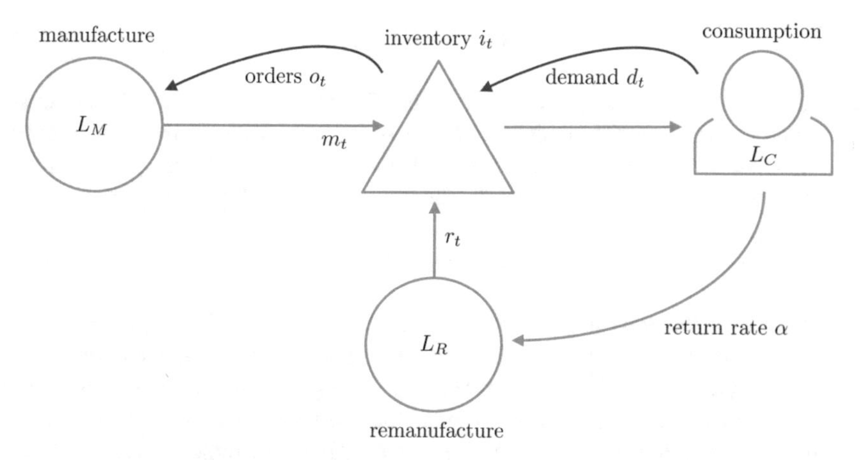

Fig. 7.1 Hybrid manufacturing/remanufacturing model

most economic option) the demand can be met by remanufactured products, so the demand deficit has to be met by new products. To do so, production orders are issued to manufacture the new products, which would be available for sales after a so-called manufacturing lead time (denoted as L_M). We also assume in the following that the realisations of L_C are higher than the realisations of L_M which, in practice, means assuming that the consumption lead time is much higher than the manufacturing lead time ($L_C >> L_M$), which is a plausible hypothesis for a wide range of products.

In this model, the sequence of events is similar to the one for the basic model discussed in Sect. 4.5 (the addition and changes are placed in italics):

1. The retailer receives from the supplier the orders placed $t - L_M$ periods before *and the remanufactured products*. For the products that have been completed, its remanufacture in period t is denoted by R_t, while the (new) units manufactured in period t are denoted by M_t. Demand is then satisfied from inventory; therefore, the following equation holds:

$$I_t = I_{t-1} - D_t + R_t + M_t$$

As usual in our models, here we also assume that negative inventory is allowed, so it is assumed that unmet demand in a period will be treated as a backlog. Note that, since the products manufactured in t have been ordered L_M periods above, we have

$$M_t = O_{t-L_M}$$

Moreover, the products remanufactured in t are a proportion of the demand that occurred L_C periods ago. This proportion α arrived at the remanufacturing facility and, after L_R periods, was available for sales, therefore:

$$R_t = \alpha \cdot D_{t-(L_R+L_C)} \tag{7.1}$$

2. The retailer studies her needs to face the future customer demand and consequently, she places a production order O_t so the new products are manufactured. As usual, we assume in principle an OUT policy where the inventory should reach a target inventory s_t. Consequently, the order has the following expression:

$$O_t = s_t - I_t - W_t \tag{7.2}$$

where W_t is the work in process as usual (orders issued that have not arrived yet). Note, however, that in this case, not only the orders issued for the manufacture of new products have to be taken into account, but also the products that have arrived at the remanufacturing facility, provided that the remanufacturing facility passes this information to the Decision Maker in charge of the replenishment. Furthermore, the Decision Maker could also guess that a fraction of the past demand will also arrive at the remanufacturing facility and that, eventually, it will be available for sales after its remanufacture, so these *future* remanufactured products could be taken into account also as work in process. Precisely, in Sect. 7.3, we will discuss the different scenarios depending on such remanufacturing pipeline visibility and on the market pipeline visibility.

7.3 Market and Remanufacturing Visibility

Different options can be considered regarding the visibility of the remanufacturing pipeline visibility and the market pipeline discussed in the previous section. Initially and as a base case, we would assume that the Decision Maker chooses to ignore all the remanufacturing flow of materials and, consequently, the work in process is solely constituted by the manufacturing orders issued in the past. In this case, it follows that

$$W_t = \sum_{i=1}^{L_M-1} O_{t-i} \tag{7.3}$$

Or equivalently,

$$W_t = W_{t-1} + O_{t-1} - M_t \tag{7.4}$$

In this case, by taking the difference in two consecutive periods in Eq. (7.2), we have

$$O_t - O_{t-1} = (s_t - s_{t-1}) - (I_t - I_{t-1}) - (W_t - W_{t-1}) \tag{7.5}$$

or

$$O_t = (s_t - s_{t-1}) + D_t - R_t \tag{7.6}$$

and substituting Eq. (7.1),

$$O_t = (s_t - s_{t-1}) + D_t - \alpha \cdot D_{t-(L_C+L_R)} \tag{7.7}$$

Assuming a time-invariant s_t and taking variances in Eq. (7.7), we have that the bullwhip effect is

$$BWE = 1 + \alpha^2 \tag{7.8}$$

Regarding the amplification of the inventory variability, if we express Eq. (7.2) in terms of inventory, and taking into account the expression of W_t in Eq. (7.3), we have that

$$I_t = s_t - O_t - \sum_{i=1}^{L_M-1} O_{t-i} = s_t - \sum_{i=0}^{L_M-1} O_{t-i} \tag{7.9}$$

If we substitute Eq. (7.7) in the previous equation, then we have

$$I_t = s_{t-L_M} - \sum_{i=0}^{L_M-1} D_{t-i} - \alpha D_{t-(L_C+L_R)-i} \tag{7.10}$$

Since the demands are iid and we have assumed that $L_C >> L_M$, it is clear that all the terms in Eq. (7.10) are independent. Therefore, taking variances, we have that

$$NSAmp = L_M(1 + \alpha^2) \tag{7.11}$$

The expressions in Eq. (7.8) and in Eq. (7.11) tell us a relatively simple relation: the amplification of the variance of the orders and the inventory increases with the return rate. In this regard, we can state (at least under the hypotheses of the model) that CLSC is subject to higher levels of bullwhip effect than its classical counterparts. The explanation lies in the additional source of variability induced by the return flow, which clearly increases as the return flow increases.

As we have discussed, we have decided to ignore the effect of the return flows in the work in process to issue the replenishment order, and this has had a negative effect on the amplification of the variability of the returns. Perhaps more sensible replenishment policies might cope with these effects. This is discussed in the next section.

7.3.1 *Estimating the Flow of the Returns*

Replenishment policies may take into account the flow of returns in the work in process. In some cases, it will be possible to have full visibility of the remanufacturing process; therefore, the Decision Maker knows for sure how many used products have arrived at the remanufacturing facility and, since these take a known L_R time units to be remanufactured, then it is possible to include these units are work in process. More specifically, the work in process W_t will be composed of the orders (to manufacture new products) launched in periods $t-1$ to $t-(L_M-1)$ (as in the base model) plus the incoming returns (i.e. all used products that have entered the remanufactured facility but are not yet completed). Since the latter are a fraction of the units sold to the customer, then it follows that

$$W_t = \sum_{i=1}^{L_M-1} O_{t-i} + \sum_{i=0}^{L_R-1} \alpha D_{t-L_C-i} \tag{7.12}$$

In this case, the difference in work in process in two consecutive periods is given by

$$W_t - W_{t-1} = O_{t-1} - O_{t-L_M} + \alpha D_{t-L_C} - \alpha D_{t-(L_C+L_R)} \tag{7.13}$$

And, in this case, Eq. (7.5) becomes

$$O_t = D_t - \alpha D_{t-L_C} \tag{7.14}$$

As we can see, taking variances, we obtain the same expression for the BWE as in the base model, i.e.

$$BWE = 1 + \alpha^2 \tag{7.15}$$

However, the amplification of the inventory variance yields a different expression, that can be obtained in a similar manner as in the previous case. In this case, Eq. (7.2) takes the following form:

$$I_t = s_t - O_t - W_t = s_t - \sum_{i=0}^{L_M-1} O_{t-i} - \sum_{i=0}^{L_R-1} \alpha D_{t-L_C-i} \tag{7.16}$$

Substituting in the above equation the expression of O_t in Eq. (7.14), we obtain

$$I_t = s_t - \sum_{i=0}^{L_M-1} D_{t-i} + \alpha \left(\sum_{i=0}^{L_M-1} D_{t-L_C-i} - \sum_{i=0}^{L_R-1} D_{t-L_C-i} \right) \tag{7.17}$$

It can be seen that the demands in the third term in the right side of Eq. (7.17) may or may not be cancelled depending on the difference between L_M and L_R. Studying the different cases of relation between L_R and L_M and taking variances, it is easy to obtain the following expression for the amplification of the inventory variance:

$$NSAmp = L_M + \alpha^2 |L_M - L_R| \tag{7.18}$$

The expression in Eq. (7.18) is a manifestation of the so-called *lead-time paradox*, which was observed in the earlier studies on CLSC, when it was observed that, in some scenarios, reducing remanufacturing lead times deteriorates the dynamic performance of the CLSC. In the model developed here, we can see that it is a consequence of the alignment (or misalignment) of the flows coming from the manufacturing and the remanufacturing process. More specifically, when both flows are aligned, the remanufacturing process produces the effect of a net subtraction in the replenishment ordering. If this is the case, the inventory amplification reaches its minimum value, equal to that obtained for the classical (open) supply chain.

7.4 Uncertainty in the Returns

In this section, we extend the models to capture the dynamic effect of the uncertainty in the returns. To do so, we study the previous model under the following different scenarios of uncertainty. In all of them, the sequence of events is the same as usual, i.e.

1. The node receives the demand D_t, the remanufactured product R_t and the new products from the supplier M_t. Inventories are then updated as follows:

$$I_t = I_{t-1} + R_t + M_t - D_t \tag{7.19}$$

2. The nodes then places an order O_t so the inventory position (on-hand inventory plus work in process) reaches a base stock level s_t:

$$O_t = s_t - I_t - W_t \tag{7.20}$$

Furthermore, we assume that the lifecycle of the product is very large as compared with the remanufacturing time, i.e. $L_C >> L_M$. Taking Eq. (7.20) and making the difference between the orders in two consecutive periods

$$O_t - O_{t-1} = (s_t - s_{t-1}) - (I_t - I_{t-1}) - (W_t - W_{t-1}) \tag{7.21}$$

As we know from Chap. 4, s_t is time-independent if the demand is idd, so $s_t = s_{t-1}$. Furthermore, assuming that the supplier has an infinite capacity, the newly manufactured product ordered in period t will arrive L_M periods later; therefore, $M_t = O_{t-L_M}$. Then, Eq. (7.21) can be written as

$$O_t - O_{t-1} = D_t - R_t - O_{t-L_M} - (W_t - W_{t-1}) \tag{7.22}$$

Equation (7.22) represents a basic expression for the orders in the closed-loop setting. Let us now introduce three scenarios regarding the uncertainty of the returns:

- **Return Volume Uncertainty (RVU)**: In this scenario, the exact volume of the return is unknown. More specifically, we assume that L_R and L_C are constant, but that the volume of the returns cannot be inferred from the amount that it was sold L_C periods before. In this case, we assume that each sold unit has a probability α to be returned (i.e. α is no longer the fixed percentage of returns but the probability that a unit is returned by the final customer). As a consequence, R_t follows a binomial distribution conditioned to the number of sold units, i.e. $R_t \sim Bin(D_{t-(L_C+L_R)}, \alpha)$. Therefore, the expected value and the variance of R_t can be obtained in the following manner:

$$E[R_t] = E[[R_t|D_t]] = E[D_t \cdot \alpha] = \alpha \cdot \mu \tag{7.23}$$

$$V[R_t] = V[E[R_t|D_t]] + E[V[R_t|D_t]] = \tag{7.24}$$
$$V[D_t \cdot \alpha] + E[D_t \cdot \alpha \cdot (1-\alpha)] = \tag{7.25}$$
$$\alpha^2\sigma^2 + \alpha \cdot (1-\alpha)\mu \tag{7.26}$$

- **Return Time Uncertainty (RTU)**: In this scenario, the exact timing of the returns is unknown. It is assumed that α is constant, but L_C is a RV with $E[L_C] = \mu_C$ and $V[L_C] = \sigma_C^2$ (Note that L_M is kept as constant). Since, in this scenario, the return ratio is constant, R_t is a function of the sales in period $t-(L_C+L_R)$: $R_t = \alpha \cdot D_{t-(L_C+L_R)}$. If we assume that the demands are iid, we have

$$V[R_t] = \alpha^2\sigma^2 \tag{7.27}$$

- **Return Quality Uncertainty (RQU)**: In this scenario, the quality of the returned product is uncertain, even if a fixed return rate α is assumed, and L_C is constant. However, L_R is assumed to be a RV with $E[L_R] = \mu_R$ and $V[L_R] = \sigma_R^2$ in order to reflect the time required to remanufacture the product, since its quality is variable. Again, R_t is a fixed proportion of the sales in time period $t-(L_C+L_R)$: $R_t = \alpha \cdot D_{t-(L_C+L_R)}$ and, assuming that the final customer demand is iid,

$$V[R_t] = \alpha^2\sigma^2 \tag{7.28}$$

In these three scenarios, the bullwhip effect can be estimated depending on whether the flow of the returns can be estimated or not. Both cases are discussed in the next subsections.

7.4.1 No Consideration of the Return Flow

If remanufactured product is not taken into account for computing the work in process, then the work in process is just the total orders placed between periods $t-1$ and $t-(L_M-1)$:

$$W_t = \sum_{i=1}^{L_M-1} O_{t-i} \tag{7.29}$$

and therefore,

$$W_t - W_{t-1} = O_{t-1} - O_{t-L_M} \quad (7.30)$$

Substituting in Eq. (7.22)

$$O_t - O_{t-1} = D_t - R_t - O_{t-L_M} - O_{t-1} + O_{t-L_M} \quad (7.31)$$

And, as a consequence,

$$O_t = D_t - R_t \quad (7.32)$$

Taking variances, since D_t is iid, we have

$$V[O_t] = V[D_t] + V[R_t] \quad (7.33)$$

So BWE has the expression

$$BWE = \frac{V[O]}{V[D]} = 1 + \frac{V[R_t]}{V[D]} \quad (7.34)$$

The specific expression for BWE would depend on $V[R_t]$, which is different for the three scenarios. In order to compute $NSAmp$, the expression of the base stock can be written in terms of the inventory, so

$$I_t = s_t - O_t - W_t \quad (7.35)$$

Since W_t is composed of the orders placed in the periods $t-1$ to $t-(L_M-1)$,

$$I_t = s_t - \sum_{i=0}^{L_M-1} O_{t-i} = s_t - \sum_{i=0}^{L_M-1} D_{t-i} - R_{t-i} \quad (7.36)$$

If we assume that $L_C >> L_M$, the terms are not correlated, so if we take variance

$$V[I_t] = \sum_{i=0}^{L_M-1} V[D_{t-i}] + V[R_{t-i}] \quad (7.37)$$

and $NSAmp$ is

$$NSAmp = L_M + \frac{\sum_{i=0}^{L_M-1} V[R_{t-i}]}{V[D]} \quad (7.38)$$

which, again, is different for the three scenarios. To determine the exact expressions of BWE and $NSAmp$, we check each scenario

- RVU: As we have seen, $V[R_t] = \alpha^2\sigma^2 + \alpha \cdot (1-\alpha)\mu$, therefore

$$Bwe = (1+\alpha^2) + \alpha \cdot (1-\alpha)\frac{\mu}{\sigma^2} \tag{7.39}$$

and

$$NSAmp = L_M \cdot \left[(1+\alpha^2) + \alpha \cdot (1-\alpha)\frac{\mu}{\sigma^2}\right] \tag{7.40}$$

- RTU and RQU: In both cases $V[R_t] = \alpha^2\sigma^2$, therefore

$$BWE = (1+\alpha^2) \tag{7.41}$$

and

$$NSAmp = L_M \cdot (1+\alpha^2) \tag{7.42}$$

7.4.2 Estimation of the Return Flow

In this case, the remanufacturer knows at time t the (remanufactured) products that will arrive in periods $t+1, \ldots, t+(L_R)$, since these are products at the end of their lifecycle that have reached its facilities. This information can be shared with the node, so the latter can take it into account to make the replenishment decisions. Here, there are two cases depending on the scenario:

- RVU and RTU: If there is no uncertainty in the remanufacturing time (L_R is constant), the information provided by the remanufacturer allows the node to know precisely the amount of remanufactured product arriving in the next L_R periods, so the node can compute its work in process as

$$W_t = \left(\sum_{i=1}^{L_M-1} O_{t-i}\right) + \left(\sum_{i=1}^{L_R} R_{t+i}\right) \tag{7.43}$$

So the differences in the work in process are now

$$W_t - W_{t-1} = \left(O_{t-1} - O_{t-L_M}\right) + \left(R_{t+L_R} - R_t\right) \tag{7.44}$$

As a consequence, substituting in Eq. (7.22),

$$O_t - O_{t-1} = D_t - R_t - O_{t-L} - O_{t-1} + O_{t-L} - R_{t+L_R} + R_t \tag{7.45}$$

or

$$O_t = D_t - R_{t+L_R} \tag{7.46}$$

- RQU: If there is uncertainty regarding the remanufacturing time, the node knows that a certain amount of product has arrived at the remanufacturing facility and that this product will eventually arrive, but the precise time cannot be determined. Therefore, it seems sensible to assume that the amount of remanufactured products would be the average number of products returned in a period ($\alpha\mu$) times the average time required to remanufacture the products (μ_R). Note that it is likely that this average remanufacturing time is not known, but that it can be estimated using, e.g. an MMSE estimation, and therefore, the mean can be used. In this case, the work in process can be computed as

$$W_t = \left(\sum_{i=1}^{L_M-1} O_{t-i}\right) + \mu_R \cdot \alpha \cdot \mu \tag{7.47}$$

So the difference in inventory is

$$W_t - W_{t-1} = \left(O_{t-1} - O_{t-L_M}\right) \tag{7.48}$$

and substituting in Eq. (7.22), we have

$$O_t = D_t - R_t \tag{7.49}$$

Both BWE and $NSAmp$ can be computed using the expressions above:

- BWE:
  - RVU scenario: Taking variances and substituting the corresponding expression of $V[R]$, we have

$$BWE = (1+\alpha^2) + \frac{\alpha(1-\alpha)\mu}{\sigma^2} \tag{7.50}$$

  - RTU and RTQ scenarios: Taking variances and substituting $V[R]$ when it is a constant ratio α of the demand,

$$BWE = (1+\alpha^2) \tag{7.51}$$

- $NSAmp$:
  - Scenarios RTU and RVU: The expression of the inventories is

$$I_t = S_t - O_t - \left(\sum_{i=1}^{(L_M-1)} O_{t-i}\right) - \left(\sum_{i=1}^{(L_R)} R_{t+i}\right) \tag{7.52}$$

  Substituting the expression of O_t according to the corresponding case,

$$I_t = S_t - \sum_{i=0}^{L_M-1} D_{t-i} + \underbrace{\sum_{i=0}^{L_M-1} R_{t+L_R-i}}_{(1)} - \underbrace{\sum_{i=1}^{(L_R)} R_{t+i}}_{(2)} \tag{7.53}$$

The term (1) takes the time periods $t + 1 + (L_R - L_M)$ to $t + L_R$, while term (2) time goes from $t + 1$ to $t + L_R$. There are different cases depending on the relationship between L_R and L_M:

1. Case $L_M = L_R$. In this case, (1) and (2) stretch over the same time periods; therefore,

$$I_t = S_t - \sum_{i=0}^{L_M-1} D_{t-i} \tag{7.54}$$

   and then it is easy to see that $NSAmp = L_M$ for both RVU and RTU.
2. Case $L_M < L_R$. In this case, (2) stretches over $(L_R - L_M)$ periods more than (1); therefore,

$$I_t = S_t - \sum_{i=0}^{L_M-1} D_{t-i} - \sum_{i=1}^{L_R-L_M} R_{t+i} \tag{7.55}$$

   and then for the RVU scenario, it can be seen that

$$NSAmp = L_M + (L_R - L_M)\left(\alpha^2 + \alpha \cdot (1 - \alpha)\frac{\mu}{\sigma^2}\right) \tag{7.56}$$

   while for the RTU scenario, we have

$$NSAmp = L_M + \alpha^2(L_R - L_M) \tag{7.57}$$

3. Case $L_M > L_R$. In this case, (1) stretches over $(L_M - L_R)$ more periods than (2); therefore, for the RVU scenario, we have

$$NSAmp = L_M + (L_M - L_R)\left(\alpha^2 + \alpha \cdot (1 - \alpha)\frac{\mu}{\sigma^2}\right) \tag{7.58}$$

   while for the RTU scenario, we have

$$NSAmp = L_M + \alpha^2(L_M - L_R) \tag{7.59}$$

As we can see, the three cases for the RVU scenario can be expressed as

$$NSAmp = L_M + |L_M - L_R|\left(\alpha^2 + \alpha \cdot (1 - \alpha)\frac{\mu}{\sigma^2}\right) \tag{7.60}$$

and for the RTU scenario as

$$NSAmp = L_M + \alpha^2 |L_M - L_R| \tag{7.61}$$

– Scenario RQU: In this case, the expression of the inventory is

$$I_t = S_t - O_t - \left(\sum_{i=1}^{(L_M-1)} O_{t-i} \right) - \mu_R \cdot \alpha \cdot \mu \tag{7.62}$$

So it is easy to check that $NSAmp = L_M$.

7.5 Conclusions

Of course, the models developed in this chapter are extremely stylised and do not take into account some of the hypotheses addressed in Chap. 6 that may not hold in certain real-life scenarios: negative orders may not be allowed or the capacity of the manufacturer may not be regarded as infinite. Furthermore, the different types of uncertainties in the returns may co-exist, which would make it extremely hard to find a close-formula expression for the amplification of the order and inventory variability. Nevertheless, these models have been useful to explain the basic issues regarding the dynamics of CLSCs (Tables 7.1 and 7.2):

- The bullwhip effect in a CLSC is no less damaging than in a traditional SC. Indeed, depending on the assumptions and if the returned product can be taken into account when making replenishment decisions, its impact in terms of the amplification of the order and inventory variability may be higher than in the traditional SC.
- In general, the possibility of taking into account the returns as work in process to make replenishment decisions improves the dynamic performance of the SC in terms of inventory, but not in terms of demand amplification.

Table 7.1 Summary of cases (BWE)

Returns	Case	BWE
No consideration of returns	Base (no unc.)	$(1+\alpha^2)$
	RVU	$(1+\alpha^2)+\alpha(1-\alpha)\frac{\mu}{\sigma^2}$
	RTU	$(1+\alpha^2)$
	RQU	$(1+\alpha^2)$
Consideration of returns	Base (no unc.)	$(1+\alpha^2)$
	RVU	$(1+\alpha^2)+\alpha(1-\alpha)\frac{\mu}{\sigma^2}$
	RTU	$(1+\alpha^2)$
	RQU	$(1+\alpha^2)$

Table 7.2 Summary of cases ($NSAmp$)

Returns	Case	$NSAmp$
No consideration of returns	Base (no unc.)	$L_M \cdot (1 + \alpha^2)$
	RVU	$L_M \cdot \left[(1 + \alpha^2) + \alpha \cdot (1 - \alpha)\frac{\mu}{\sigma^2}\right]$
	RTU	$L_M \cdot (1 + \alpha^2)$
	RQU	$L_M \cdot (1 + \alpha^2)$
Consideration of returns	Base (no unc.)	$L_M + \alpha^2 \cdot \lvert L_R - L_M \rvert$
	RVU	$L_M + (\alpha^2 + \frac{\alpha(1-\alpha)}{\sigma^2}\mu) \cdot \lvert L_R - L_M \rvert$
	RTU	L_M
	RQU	L_M

- If the returns are taken into account, there exists a lead-time paradox: the more similar the remanufacturing lead time and the (new products) lead time are, the lesser the bullwhip effect in terms of inventory amplification.
- The influence of the different sources of uncertainty in the returns is diverse: while the uncertainty in the volume of returned products increases the bullwhip effect both in terms of demand and inventory amplification, other sources of uncertainty (such as uncertainty in the quality or timing of the returns) do not necessarily increase the bullwhip effect.

7.6 Further Readings

In [1, 2], SCs are mentioned as the main source of carbon in the environment that lead to global warming and climate change. Some figures regarding the sales of remanufactured products are given in [3, 4]. There are several contribution reviewing different aspects of the literature on CLSC, including [5–11]. Reference [12] reviews different decisions in hybrid manufacturing/remanufacturing systems. The hybrid manufacturing/remanufacturing model in this chapter was first described by [13] and it was subsequently adopted by the majority of the literature (see, e.g. [14, 15] or [16]). Empirical data supporting the aforementioned hybrid model can be found in [17] or in [18].

The derivation of the expressions of BWE and $NSAmp$ has been obtained from [19]. Some conflicting results to the ones reported in the latter references (although obtained using different hypotheses) regarding the bullwhip effect in CLSCs can be found in [20–22] or in [23]. The lead-time paradox was first observed by [24]. Further research on this effect include [13, 16]. The developments in Sect. 7.4 have been taken from [25]. The uncertainties in the return yield have been addressed by [17] using a control systems approach and in [19] using simulation. Further analysis on the effect of quality grading of the returns is presented in [26].

The area of CLSC dynamics keeps growing and it is being enriched with different contributions. A systematic review of the developments in the area can be found in [9]. The use of proportional controllers to mitigate the bullwhip effect in a CLSC has been investigated by different researchers, including [27]. The value of information sharing in the CLSC context is analysed by [28]. The intersection of CLSC and ADI is explored in [29]. The effect of the remanufacturing configuration in the dynamic performance of a SC is studied in [30], and the bullwhip effect in a CLSC with capacity limits is analysed in [31]. The term *green bullwhip effect* has been coined by [32] to denote the amplification of the changes in environmental requirements that are passed along the SC. Finally, other contributions in the field of CLSC include [33–35] or [36].

References

1. Ahmed, W., Sarkar, B.: Impact of carbon emissions in a sustainable supply chain management for a second generation biofuel. J. Clean. Prod. **186**, 807–820 (2018)
2. Ahmed, W., Sarkar, B.: Management of next-generation energy using a triple bottom line approach under a supply chain framework. Resour. Conserv. Recycl. **150** (2019)
3. Abbey, J., Guide V.D.R., J.: A typology of remanufacturing in closed-loop supply chains. Int. J. Prod. Res. **56**(1–2), 374–384 (2018)
4. Abbey, J., Guide V.D.R., J.: Closed-loop supply chains: a strategic overview. In: Sustainable Supply Chains: a Research-based Textbook on Operations and Strategy, pp. 375–393 (2017)
5. Guide, V., Jr., Van Wassenhove, L.: The evolution of closed-loop supply chain research. Oper. Res. **57**(1), 10–18 (2009)
6. Souza, G.: Closed-loop supply chains: a critical review, and future research*. Decis. Sci. **44**(1), 7–38 (2013)
7. Schenkel, M., Caniëls, M., Krikke, H., Van Der Laan, E.: Understanding value creation in closed loop supply chains - past findings and future directions. J. Manuf. Syst. **37**, 729–745 (2015)
8. Govindan, K., Soleimani, H., Kannan, D.: Reverse logistics and closed-loop supply chain: a comprehensive review to explore the future. Eur. J. Oper. Res. **240**(3), 603–626 (2015)
9. Braz, A., De Mello, A., de Vasconcelos Gomes, L., de Souza Nascimento, P.: The bullwhip effect in closed-loop supply chains: a systematic literature review. J. Clean. Prod. **202**, 376–389 (2018)
10. Cannella, S., Bruccoleri, M., Framinan, J.: Closed-loop supply chains: what reverse logistics factors influence performance? Int. J. Prod. Econ. **175**, 35–49 (2016)
11. Goltsos, T., Ponte, B., Wang, S., Liu, Y., Naim, M., Syntetos, A.: The boomerang returns? accounting for the impact of uncertainties on the dynamics of remanufacturing systems. Int. J. Prod. Res. **57**(23), 7361–7394 (2019)
12. Aras, N., Verter, V., Boyaci, T.: Coordination and priority decisions in hybrid manufacturing/remanufacturing systems. Prod. Oper. Manag. **15**(4), 528–543 (2006)
13. Tang, O., Naim, M.: The impact of information transparency on the dynamic behaviour of a hybrid manufacturing/remanufacturing system. Int. J. Prod. Res. **42**(19), 4135–4152 (2004)
14. Hosoda, T., Disney, S., Gavirneni, S.: The impact of information sharing, random yield, correlation, and lead times in closed loop supply chains. Eur. J. Oper. Res. **246**(3), 827–836 (2015)

15. Ponte, B., Naim, M., Syntetos, A.: The value of regulating returns for enhancing the dynamic behaviour of hybrid manufacturing-remanufacturing systems. Eur. J. Oper. Res. **278**(2), 629–645 (2019)
16. Hosoda, T., Disney, S.: A unified theory of the dynamics of closed-loop supply chains. Eur. J. Oper. Res. **269**(1), 313–326 (2018)
17. Zhou, L., Naim, M., Disney, S.: The impact of product returns and remanufacturing uncertainties on the dynamic performance of a multi-echelon closed-loop supply chain. Int. J. Prod. Econ. **183**, 487–502 (2017)
18. Zhang, S., Li, X., Zhang, C.: A fuzzy control model for restraint of bullwhip effect in uncertain closed-loop supply chain with hybrid recycling channels. IEEE Trans. Fuzzy Syst. **25**(2), 475–482 (2017)
19. Ponte, B., Framinan, J., Cannella, S., Dominguez, R.: Quantifying the bullwhip effect in closed-loop supply chains: the interplay of information transparencies, return rates, and lead times. Int. J. Prod. Econ. **230** (2020)
20. Adenso-Díaz, B., Moreno, P., Gutiérrez, E., Lozano, S.: An analysis of the main factors affecting bullwhip in reverse supply chains. Int. J. Prod. Econ. **135**(2), 917–928 (2012)
21. Corum, A., Vayvay, O., Bayraktar, E.: The impact of remanufacturing on total inventory cost and order variance. J. Clean. Prod. **85**, 442–452 (2014)
22. Ma, L., Chai, Y., Zhang, Y., Zheng, L.: Modeling and analysis of the bullwhip effect in remanufacturing closed-loop supply chain. Appl. Mech. Mater. **541–542**, 1556–1561 (2014)
23. Zhou, L., Naim, M., Tang, O., Towill, D.: Dynamic performance of a hybrid inventory system with a kanban policy in remanufacturing process. Omega **34**(6), 585–598 (2006)
24. Inderfurth, K., Van Der Laan, E.: Leadtime effects and policy improvement for stochastic inventory control with remanufacturing. Int. J. Prod. Econ. **71**(1–3), 381–390 (2001)
25. Framinan, J., Cannella, S., Dominguez, R.: Addressing uncertainty variability in closed-loop supply chains. Technical Report TROI-2020-03, Industrial Management Research Group (2020)
26. Ponte, B., Cannella, S., Dominguez, R., Naim, M., Syntetos, A.: Quality grading of returns and the dynamics of remanufacturing. Int. J. Prod. Econ. **236** (2021)
27. Cannella, S., Ponte, B., Dominguez, R., Framinan, J.: Proportional order-up-to policies for closed-loop supply chains: the dynamic effects of inventory controllers. Int. J. Prod. Res. (2021)
28. Dominguez, R., Ponte, B., Cannella, S., Framinan, J.: Building resilience in closed-loop supply chains through information-sharing mechanisms. Sustainability (Switzerland) **11**(23) (2019)
29. Chen, L., Yücel, S., Zhu, K.: Inventory management in a closed-loop supply chain with advance demand information. Oper. Res. Lett. **45**(2), 175–180 (2017)
30. Dominguez, R., Cannella, S., Framinan, J.: Remanufacturing configuration in complex supply chains. Omega (United Kingdom) **101** (2021)
31. Tombido, L., Louw, L., van Eeden, J., Zailani, S.: A system dynamics model for the impact of capacity limits on the bullwhip effect (bwe) in a closed-loop system with remanufacturing. J. Remanuf. (2021)
32. Lee, S.Y., Klassen, R., Furlan, A., Vinelli, A.: The green bullwhip effect: transferring environmental requirements along a supply chain. Int. J. Prod. Econ. **156**, 39–51 (2014)
33. Zhao, Y., Cao, Y., Li, H., Wang, S., Liu, Y., Li, Y., Zhang, Y.: Bullwhip effect mitigation of green supply chain optimization in electronics industry. J. Clean. Prod. **180**, 888–912 (2018)
34. Tombido, L., Louw, L., van Eeden, J.: The bullwhip effect in closed-loop supply chains: a comparison of series and divergent networks. J. Remanuf. **10**(3), 207–238 (2020)
35. Tombido, L., Baihaqi, I.: The impact of a substitution policy on the bullwhip effect in a closed loop supply chain with remanufacturing. J. Remanuf. **10**(3), 177–205 (2020)
36. Sepulveda-Rojas, J., Ternero, R.: Analysis of the value of information and coordination in a dyadic closed loop supply chain. Sustainability (Switzerland) **12**(20), 1–18 (2020)

Chapter 8
Modelling Complex SC Structures

8.1 Introduction

In the models developed in the preceding chapters of the book, we assume that the SC has a linear structure. This may not be the case in most real-life SCs, where complex structures with several nodes in an echelon may appear. In these complex SC structures, the challenge is to evaluate the dynamic effects of the order and inventory amplifications. Clearly, given the complexity of the computations, the methodology of choice would be to build simulation models of the SC, which would be specific and where little generalizable results can be gained, therefore, we will not discuss these in this chapter and instead, we will focus on some general insights on the bullwhip effect caused by several factors which are inherent to the structure of the SC.

In some cases, it is possible to transform the non-linear structure into a linear one by substituting the different nodes in an echelon with an equivalent node, which would make the resulting SC tractable from an analytical point of view. However, this approximation does not always result in an equivalent dynamic performance, as there are additional factors influencing the bullwhip effect. As we will see in this chapter, some of these factors include the fact that there are additional amplifications in the variability that may be caused by the manner in which several nodes in an echelon ordering batches to a previous echelon, or what happen if these nodes compete for the same customer (which is located in the next echelon in the SC). More specifically, in this chapter, we

- Discuss how having several retailers ordering batches affects the bullwhip effect depending on how these orders take place (Sect. 8.2). In this case, it is assumed that each retailer has its own demand, i.e. do not compete for the demand downstream in the SC.
- Discuss how the competition for the demand among several retailers may affect the bullwhip effect, even without considering batches (8.3).

J. M. Framinan, *Modelling Supply Chain Dynamics*,
https://doi.org/10.1007/978-3-030-79189-6_8

8.2 Several Retailers/Order Batching

Perhaps the first documented proof of the bullwhip effect caused by a non-linear structure is the case of several retailers with order batching. In this model, we consider a part of an SC where a supplier must face the demand of n retailers, which, in turn, face the demand of the final customer. We assume that the retailers are identical and they review their inventory every r period, placing an order for a quantity ξ_{jt}. Figure 8.1 depicts the structure of the SC.

Therefore, the demand faced by retailer j in time period t is ξ_{jt}, and it is assumed that their expected value is $E[\xi_{jt}] = \mu$ and that its variance is $V[\xi_{jt}] = \sigma^2$. Furthermore, the demands ξ_{jt} are assumed to be mutually iid. As a consequence, D_t the total demand experienced in the retailer echelon in period t would be

$$D_t = \sum_{j=1} \xi_{jt} \tag{8.1}$$

Clearly, as we assume that the demands are iid, the expressions for $E[D_t]$ and $V[D_t]$ are straightforward

$$E[D_t] = n \cdot \mu \tag{8.2}$$

and

$$V[D_t] = n \cdot \sigma^2 \tag{8.3}$$

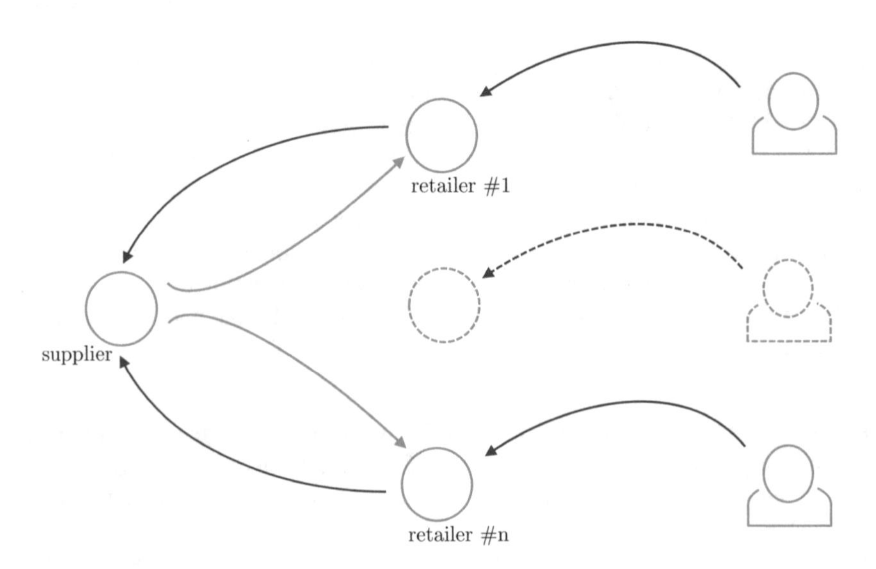

Fig. 8.1 SC with several retailers ordering batches

Let us assume that the final customers are willing to wait if the demand cannot be met (full backlog) and that the retailers use an OUT replenishment policy. This means that every time one retailer issues an order, the quantity ordered would be simply the demand that he/she has experienced since the last order was issued. More specifically, if retailer j places an order in time period t, then he/she would place an order to the supplier equal to O_{jt}, where

$$O_{jt} = \sum_{k=t-(r-1)}^{t} \xi_{jt} \tag{8.4}$$

(note again that we assume that, as usual, at the time t where the retailer j places the order, he/she already knows the demand in period t and this demand has been already satisfied. Therefore, the summation index goes from $r - 1$ to 0)

Let us denote by O_t the total demand received by the supplier in period t. Such total demand would be then the sum of all O_{jt} different than zero, i.e. the sum of the orders from the retailers that place an order in period t. Since the retailers are identical, we can write

$$O_t = \sum_{j=1}^{N_t} \sum_{k=t-(r-1)}^{t} \xi_{jt} \tag{8.5}$$

where N_t is the total number of retailers placing an order in period t. Note that N_t is a RV that depends on whether the retailers place their orders independently, or if there is some sort of organisation among the retailers (either naturally or induced by the supplier) to place the orders.

Clearly, to measure the bullwhip effect, we intend to check the ratio $V[O_t]/V[D_t]$, for which we need to determine the variance of O_t. Since O_t depends on N_t, we need to make some assumptions regarding how the retailers place their orders. More specifically, we will study the following cases:

- Independent retailers (case 1): Each retailer places its order independently from the others.
- Positively correlated orders (case 2): All retailers place their orders in the same period.
- Balanced orders (case 3): The number of retailers placing an order in the periods is balanced.

These cases are developed in the next subsections.

8.2.1 Case 1: Independent Retailers

If the retailers are independent, but all of them have a replenishment cycle of r periods, clearly, the probability for a retailer to place an order in a given period is

$\frac{1}{r}$ (and $1 - \frac{1}{r}$ the probability of not placing an order). In order words, a Bernoulli distribution with success probability $\frac{1}{r}$ yields one if the retailer places the order, and zero otherwise. Then it follows that N_t the number of retailers placing an order in a given period t follows a binomial distribution with parameters n and $\frac{1}{r}$, i.e. $N_t \sim Bin(n, \frac{1}{r})$.

Under this assumption, let us compute $E[O_t]$ and $V[O_t]$. Regarding $E[O_t]$, we can apply the law of the total mean, i.e.

$$E[O_t] = E[E[O_t|N_t]] \tag{8.6}$$

i.e. we first compute the conditional expected value of O_t assuming that N_t is constant, and then we compute the expected value (with respect to N_t) of the resulting expression. More specifically

$$E[O_t|N_t] = E\left[\sum_{j=1}^{N_t} \sum_{k=t-(r-1)}^{t} \xi_{jt}\right] = r \cdot N_t \cdot \mu \tag{8.7}$$

and then

$$E[O_t] = E[E[O_t|N_t]] = r \cdot n\frac{1}{r} \cdot \mu = n \cdot \mu \tag{8.8}$$

The so-obtained value is foreseeable, as the mean size of the orders received by the supplier should be equal to the mean demand of the final customer across the suppliers.

Regarding $V[O_t]$, we know that, according to the total variance

$$V[O_t] = E[V[O_t|N_t]] + V[E[O_t|N_t]] \tag{8.9}$$

Computing the first of the two terms and taking into account that ξ_{jt} are iid, we have

$$V[O_t|N_t] = V[\sum_{j=1}^{N_t} \sum_{k=t-(r-1)}^{t} \xi_{jt}|N_t] = \sum_{j=1}^{N_t} \sum_{k=t-(r-1)}^{t} V[\xi_{jt}] = N_t \cdot r \cdot \sigma^2 \tag{8.10}$$

and

$$E[V[O_t|N_t]] = r \cdot \sigma^2 \cdot E[N_t] = n \cdot \sigma^2 \tag{8.11}$$

Regarding the second term of Eq. (8.9), and plugging Eq. (8.7) into this second term

$$V[E[O_t|N_t]] = V[r \cdot N_t \cdot \mu] = r^2 \cdot \mu^2 \cdot n\frac{1}{r}(1 - \frac{1}{r}) = \mu^2 \cdot n \cdot (r - 1) \tag{8.12}$$

As a result, the expression of $V[O_t]$ is

$$V[O_t] = n \cdot \sigma^2 + \mu^2 \cdot n \cdot (r - 1) \tag{8.13}$$

To obtain the BWE, we take into account that $V[D_t] = n \cdot \sigma^2$

$$BWE = 1 + \left(\frac{\mu}{\sigma}\right)^2 (r - 1) \tag{8.14}$$

As we can see, assuming that $\sigma \neq 0$ (otherwise the BWE indicator makes no sense), the bullwhip effect exists ($BWE > 1$) for $r > 1$, i.e. if the review period is higher than one period. It increases with the length of the review period, and also with the inverse of the coefficient of variation of the demand, i.e. whenever the variability of the demand is higher (measured as the coefficient of variation of the demand), then the bullwhip effect diminishes. Note that BWE measures the amplification of the variance of the demand across the SC, therefore, this result should be interpreted in the sense that a higher *input* variability is not greatly amplified. Finally, it is interesting to note that the bullwhip effect does not depend on the number of retailers as long as they are independent.

8.2.2 *Case 2: Positively Correlated Orders*

In this case, all retailers place their orders in the same period t, i.e. they place an order every r period, but the time period chosen among this r is the same for all n retailers. Although this situation is somehow academic, it could serve to model the case where all (or at least most) retailers place their orders at the end (or beginning) of the week or month, thus inducing a high fluctuation in the demand received by the supplier.

As in the previous case, the goal is to find an expression for $V[O_t]$, for which we have Eq. (8.9). In this case, N_t can only take two values: 0 if t is a period where none of the retailers places an order, and n if t is the period where all the retailers place their orders. Clearly, $P[N_t = n] = \frac{1}{r}$ whereas $P[N_t = 0] = 1 - \frac{1}{r}$.

As a consequence, the expected value and the variance of N_t are now the following:

$$E[N_t] = \frac{n}{r} \tag{8.15}$$

and

$$V[N_t] = \frac{n^2}{r}\left(1 - \frac{1}{r}\right) \tag{8.16}$$

Note that the expected value is the same as in the previous case (which makes sense, as the needs for replenishment are the same), but the variance is n times higher (the variance of the binomial in the previous case is $\frac{n}{r}(1 - \frac{1}{r})$).

Using Eq. (8.9), we have that the first term is now

$$E[V[O_t|N_t]] = E[\sum_{j=1}^{N_t} \sum_{k=t-(r-1)}^{t} V[\xi_{jt}]|N_t] = E[N_t \cdot r \cdot \sigma^2] = n \cdot \sigma^2 \quad (8.17)$$

while the second term is

$$V[E[O_t|N_t]] = V[r \cdot N_t \cdot \mu] = r^2 \cdot \mu^2 \cdot n^2 \frac{1}{r}(1 - \frac{1}{r}) = \mu^2 \cdot n^2 \cdot (r-1) \quad (8.18)$$

As a result, the variance in this case is

$$V[O_t] = n \cdot \sigma^2 + \mu^2 \cdot n^2 \cdot (r-1) \quad (8.19)$$

and the bullwhip effect is

$$BWE = 1 + \left(\frac{\mu}{\sigma}\right)^2 n(r-1) \quad (8.20)$$

Note that, as in the previous case, the bullwhip effect is always higher than 1 if $r > 1$. However, in this case, the number of retailers positively affects the bullwhip effect in a proportional manner. This is particularly interesting, as, in some situations, the correlation of the orders appears 'naturally' due to the operational customs of the retailers (such as, e.g. ordering at the end of the week to start over the new week with fresh inventory.)

8.2.2.1 Case 3: Balanced Orders

In this case, let us assume that the number of orders in each period over the replenishment cycle r is (approximately) equally divided among the n retailers. More specifically, let us assume that it is possible to assign m retailers to each period and that each one of the reminder k retailers (obviously $k < r$) is assigned to a different period. As a result, we have k periods with orders from $m + 1$ retailers, and $r - k$ periods with orders from m retailers. Note that this is the most balanced allocation of the retailers that we can do and, even if it may seem a bit unnatural, this case may be *forced* in some cases by the supplier, using different time windows for the retailers to place their orders.

As a result of this case, we have that $n = m \cdot r + k$ with $k < r$, k periods with $m + 1$ retailers and $r - k$ periods with m retailers. In this case, N_t the number of retailers that place an order in a given period t may be $m + 1$ (with probability k/r) or m (with probability $1 - \frac{k}{r}$). We can then compute the mean and variance of N_t

$$E[N_t] = \frac{n}{r} \quad (8.21)$$

and

$$V[N_t] = \frac{k}{r}\left(1 - \frac{k}{r}\right) \tag{8.22}$$

Using Eq. (8.9), we have that the first term is now

$$E[V[O_t|N_t]] = E[\sum_{j=1}^{N_t} \sum_{k=t-(r-1)}^{t} V[\xi_{jt}]|N_t] = E[N_t \cdot r \cdot \sigma^2] = n \cdot \sigma^2 \tag{8.23}$$

while the second term is

$$V[E[O_t|N_t]] = V[r \cdot N_t \cdot \mu] = r^2 \cdot \mu^2 \cdot \frac{k}{r}\left(1 - \frac{k}{r}\right) = \mu^2 \cdot k \cdot (r - k) \tag{8.24}$$

As a result, the variance in this case is

$$V[O_t] = n \cdot \sigma^2 + \mu^2 \cdot k \cdot (r - k) \tag{8.25}$$

and the bullwhip effect is

$$BWE = 1 + \left(\frac{\mu}{\sigma}\right)^2 k(r - k) \tag{8.26}$$

Note that, differently from the other cases, there is no amplification of the variance of the demand if the retailers are perfectly balanced among the r periods (i.e. if $k = 0$). The imbalance caused by the impossibility of allocating the same number of retailers to the periods increases the bullwhip effect in a symmetric manner. Also, interestingly, in this case, there is no influence on the number of retailers as long as they are balanced among the periods.

8.2.3 Comparing the Cases

Table 8.1 shows the different values of the indicator BWE for the three cases discussed in the previous sections.

Table 8.1 Expressions of the bullwhip effect depending on the correlation of the retailers

Case	BWE
Independent retailers	$1 + \left(\frac{\mu}{\sigma}\right)^2 (r - 1)$
Positively correlated retailers	$1 + \left(\frac{\mu}{\sigma}\right)^2 n(r - 1)$
Balanced retailers	$1 + \left(\frac{\mu}{\sigma}\right)^2 k(r - k)$

As it can be seen in Table 8.1, unless the retailers are perfectly balanced, the replenishment cycle r plays the role of exacerbating the bullwhip effect. However, the number of retailers only increases the bullwhip effect if their orders are positively correlated.

Several managerial insights can be obtained by looking at these results:

- It is only possible to remove the bullwhip effect if the orders of the retailers are perfectly balanced. Therefore, if it is possible, the main goal of the supplier is to balance such orders. This may be done by allocating different time windows for their customers (the retailers) to place their orders, perhaps using economic incentives (discounts) to foster the balance of the orders.
- If it is not possible to balance the orders of the different retailers, then the replenishment period r has a linear influence on the bullwhip effect. Since r it is usually the result of the optimal trade-off between the (fixed) order costs and the (variable) inventory costs (see e.g. the EOQ formula), then the natural way to reduce r would be to reduce these fixed costs by, e.g. reducing order processing costs by introducing automated order processing.[1] An additional option would be to reduce the fixed transportation costs by, e.g. outsourcing the transportation activities. However, note that these actions should be taken by the retailer who does not suffer the bullwhip effect, therefore, it is the supplier who must foster the reduction of the replenishment cycle by not favouring high-volume, low-frequency orders, e.g. by not offering volume discounts. Additionally, if the supplier provides different products to the retailer, he/she can offer such volume discounts referred to the whole order (including the different products) and not taking into account the volume of a specific product.
- The number of retailers also has an influence on the bullwhip effect if these retailers are not independent or balanced among the replenishment cycle. In other words, the effect of the correlation among retailers on the bullwhip effect increases with the number of retailers.

8.3 Several Retailers with a Competing Market

In this section, we address the case of an SC where two retailers share the final customer demand. The case is depicted in Fig. 8.2. More specifically, we assume that there are two retailers sharing the total demand D_t of a final customer which, as usual, is modelled using an AR(1)

$$D_t = d + \rho \cdot D_{t-1} + \epsilon_t \tag{8.27}$$

[1] In some studies, it has been found that automated order processing may reduce the order processing costs by one tenth with respect to the manual processing costs.

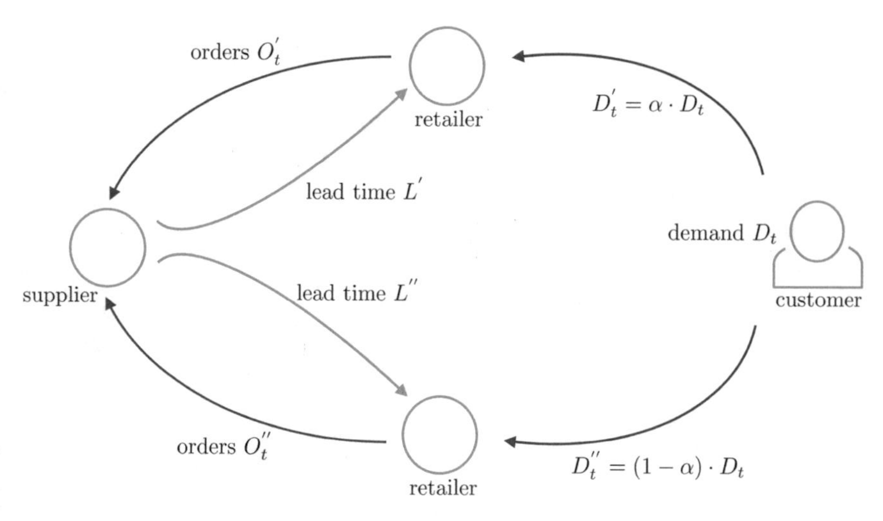

Fig. 8.2 SC with two retailers sharing a fixed customer demand

where $|\rho| < 1$ and ϵ_t are iids with $\epsilon_t \sim N(0, \sigma^2)$. Each retailer has a market share α which is assumed to be fixed, so at time period t the demand faced by Retailer #1 is $D_t' = \alpha D_t$, whereas the demand faced by Retailer #2 is $D_t'' = (1 - \alpha) D_t$.

Furthermore, we assume that both the retailers operate under an MTS strategy, and therefore, build an inventory to face their demand. More specifically, they use an OUT policy (see Sect. 2.4) to replenish their inventories, so each period they place orders (denoted as O_t' and O_t'') to their supplier. In general, we assume that the lead time for each retailer is constant, but different for each retailer (L' and L'').

Clearly, the sum of the orders of the two retailers is the demand experienced by the supplier (denoted as O_t), i.e. $O_t = O_t' + O_t''$. We are interested to find an expression for the variance amplification experienced by the supplier, measured as $BWE = \frac{V[O_t]}{V[D]}$. The interest of finding the expression for BWE, in this case, is to analyse the dynamic effects induced by having several retailers with different lead times. Furthermore, it can be interesting to compare such amplification of the order variance with that of an alternative scenario where the supplier satisfies the orders placed by the retailers using a third-party warehouse such as the one depicted in Fig. 8.3. In this manner, we can see that, if we assume that the lead time from the warehouse to the retailers is negligible, then this last situation is equivalent to the case where the supplier has only one retailer. By comparing the BWE in both cases, we can gain some managerial insights on the requirements to advantageously use third-party warehouses.

In order to develop the mathematical formulae, let us start by stating that, if D_t is AR(1), then both D_t' and D_t'' are also AR(1) with the following parameters:

- $D_t' = d' + \rho \cdot D_{t-1}' + \epsilon_t'$ with $d' = \alpha d$ and $\epsilon_t' \sim N\left(0, (\alpha\sigma)^2\right)$
- $D_t'' = d'' + \rho \cdot D_{t-1}'' + \epsilon_t''$ with $d'' = (1-\alpha)d$ and $\epsilon_t' \sim N\left(0, ((1-\alpha)\sigma)^2\right)$

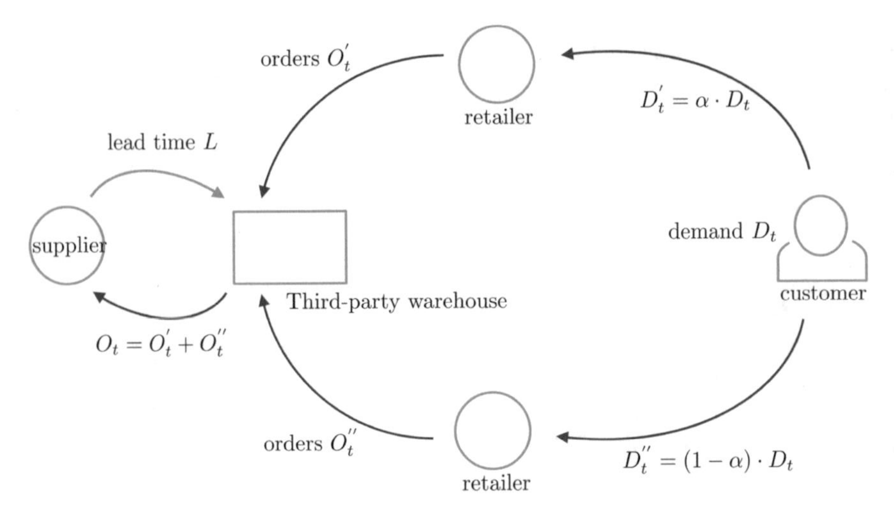

Fig. 8.3 SC with two retailers and a third-party warehouse for the retailers

The above can be seen in an easier manner if we write, e.g. $D_t^{'}$ as a function of the causal form of D_t (see Eq. (2.2)), i.e.

$$D_t^{'} = \alpha D_t = \alpha \left(\frac{d}{1-\rho} + \sum_{i=0}^{\infty} \rho^i \cdot \epsilon_{t-i} \right) = \frac{(\alpha \cdot d)}{1-\rho} + \sum_{i=0}^{\infty} \rho^i \cdot (\alpha \cdot \epsilon_{t-i}) \tag{8.28}$$

Since Retailer #1 adopts an OUT policy and assuming that the excess can be returned without additional costs, $O_t^{'}$ the order placed to the supplier in period t would be (see Eq. (4.15))

$$O_t^{'} = s_t^{'} - s_{t-1}^{'} + D_t^{'} \tag{8.29}$$

where $s_t^{'}$ is the base stock level in period t, which we know (see Eq. (2.55)) is:

$$s_t^{'} = \hat{d}_t^{L^{'}} + z \cdot \hat{\sigma}_t^{L^{'}} \tag{8.30}$$

where $\hat{d}_t^{L^{'}}$ is the estimate of the retailer's demand across $L^{'}$ periods and $\hat{\sigma}_t^{L^{'}}$ the standard deviation of the error of such estimation. As a consequence, $O_t^{'}$ can be written

$$O_t^{'} = (\hat{d}_t^{L^{'}} - \hat{d}_{t-1}^{L^{'}}) + z \cdot \left(\hat{\sigma}_t^{L^{'}} - \hat{\sigma}_{t-1}^{L^{'}} \right) + D_t^{'} \tag{8.31}$$

If we assume that Retailer #1 performs an MMSE estimate of the demand, then we know that the standard deviation of the error would not depend on t, therefore, the above expression can be simplified as

$$O_t^{'} = \hat{d}_t^{L^{'}} - \hat{d}_{t-1}^{L^{'}} + D_t^{'} \tag{8.32}$$

Furthermore, taking into account that $D_t^{'}$ is an AR(1) with the parameters above discussed, and also from Eq. (4.20) that the expression of the $\hat{d}_t^{L'}$ when an MMSE estimation is performed, we have

$$\hat{d}_t^{L'} = L' \frac{\alpha d}{1-\rho} + \rho \frac{1-\rho^{L'}}{1-\rho} \left(D_t^{'} - \frac{\alpha \cdot d}{1-\rho} \right) \tag{8.33}$$

As a result, we have

$$O_t^{'} = D_t^{'} + \rho \frac{1-\rho^{L'}}{1-\rho} (D_t^{'} - D_{t-1}^{'}) = (1+K^{'}) \cdot D_t^{'} - K^{'} \cdot D_{t-1}^{'} = \tag{8.34}$$

$$\alpha(1+K^{'}) \cdot D_t - \alpha K^{'} \cdot D_{t-1} \tag{8.35}$$

where we define $K^{'}$ as follows:

$$K^{'} = \sum_{i=1}^{L^{'}} \rho^i = \rho \frac{1-\rho^{L'}}{1-\rho} \tag{8.36}$$

Note that this definition of $K^{'}$ is equivalent to that in Eq. (4.23) in Chap. 4 for the case of one retailer and identical hypotheses (return of negative orders, MMSE estimation, OUT policy, etc.).

Following the same logic for Retailer #2, it is easy to see that the order placed in period t is

$$O_t^{''} = (1-\alpha)(1+K^{''}) \cdot D_t - (1-\alpha)K^{''} \cdot D_{t-1} \tag{8.37}$$

with

$$K^{''} = \sum_{i=1}^{L^{''}} \rho^i = \rho \frac{1-\rho^{L''}}{1-\rho} \tag{8.38}$$

The sum of $O_t^{'}$ and $O_t^{''}$ is the total demand experienced by the supplier, i.e.

$$O_t = O_t^{'} + O_t^{''} = [\alpha(1+K^{'}) + (1-\alpha)(1+K^{''})] \cdot D_t - [\alpha K^{'} + (1-\alpha)K^{''}] \cdot D_{t-1} \tag{8.39}$$

In order to abbreviate the expression, we can define $\mathcal{K}$ as

$$\mathcal{K} = \alpha K^{'} + (1-\alpha)K^{''} \tag{8.40}$$

So the expression of the demand faced by the supplier is

$$O_t = (1 + \mathcal{K}) \cdot D_t + \mathcal{K} \cdot D_{t-1} \tag{8.41}$$

Taking variance, we have

$$V[O_t] = (1 + \mathcal{K})^2 \sigma^2 + \mathcal{K}\sigma^2 - 2\rho (1 + \mathcal{K}) \cdot \mathcal{K}\sigma^2 \tag{8.42}$$

Therefore, the bullwhip effect is

$$BWE = 1 + 2\mathcal{K}(1 + \mathcal{K})(1 - \rho) \tag{8.43}$$

Note the parallelism between BWE in Eq. (8.43) and the expression of BWE in Eq. (4.26) of Chap. 4, which corresponds to the bullwhip effect experienced by the supplier if there is one retailer in the SC. Both expressions are very similar, being the only difference that, for the case of the two retailers in a competing market, we have a factor $\mathcal{K}$ which is a linear combination of the individual K of each retailer. Therefore, there is a relatively easy way to compare the bullwhip effect experienced by the supplier when there is only one retailer (recall that this can be used to model the usage of a third-party warehouse, thus in the following is denoted as BWE_{3PW}), and the case where there are two retailers competing for the same market with a fixed share (in the following denoted as BWE_{2R}).

More specifically, there are some special cases that we can treat in an easy manner

- If $\rho = 0$, then $BWE_{3PW} = BWE_{2R}$. This is easy to see by checking that, if $\rho = 0$, then both $K^{'}$ and $K^{''}$ are equal to zero, and consequently, $\mathcal{K} = 0$ for any value of α. Therefore, we can state that, if the customer demand is iid, then the bullwhip effect cannot be diminished by employing a third-party warehouse.
- If $L^{'} = L_{''} = L$, then $BWE_{3PW} = BWE_{2R}$. Again, this can be easily proved by taking into account that, if $L^{'} = L_{''}$, then $K^{'} = K^{''} = K$ and, as a consequence, $\mathcal{K} = K$. Therefore, if the lead time of the retailers is the same, then there are no differences with respect to BWE in both cases regardless the values of ρ (coefficient of correlation) or α (retailers' market share).

Other cases can be more complex, although it is relatively easy to show that BWE_{2R} is higher than one if and only if $\rho > 0$. Nevertheless, the proof is quite dull in terms of algebra and it will be omitted here.

In order to numerically analyse the difference between the two cases, we can plot the bullwhip effect indicator depending on the correlation and for different values of $L^{''}$ ($L^{'} = 1$ for all cases). This is shown in Fig. 8.4. In Fig. 8.5, we show BWE_{2R} as a function of α for different values of $L^{'}$ ($L^{''} = 1$ in all curves). As it can be seen, the bullwhip effect increases (at least in this case) with $L^{''}$.

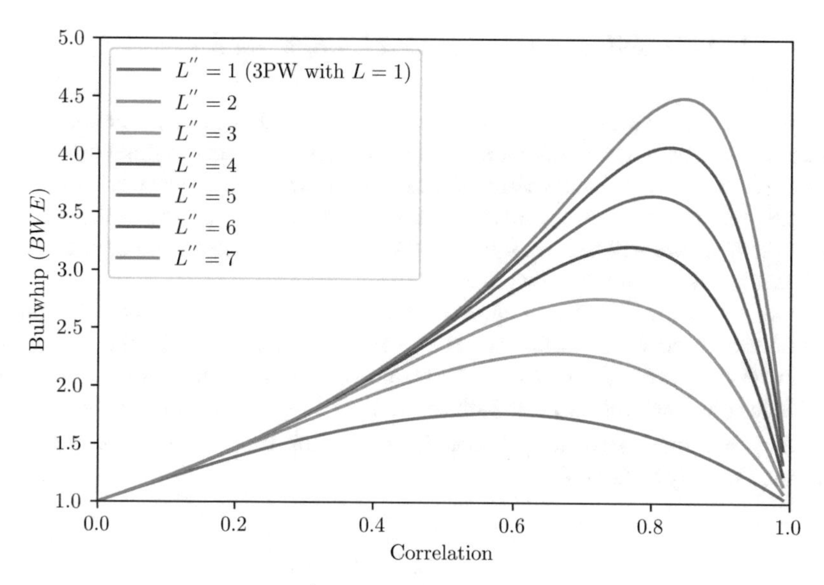

Fig. 8.4 BWE depending on ρ for $L' = 1$ and $\alpha = 0.3$

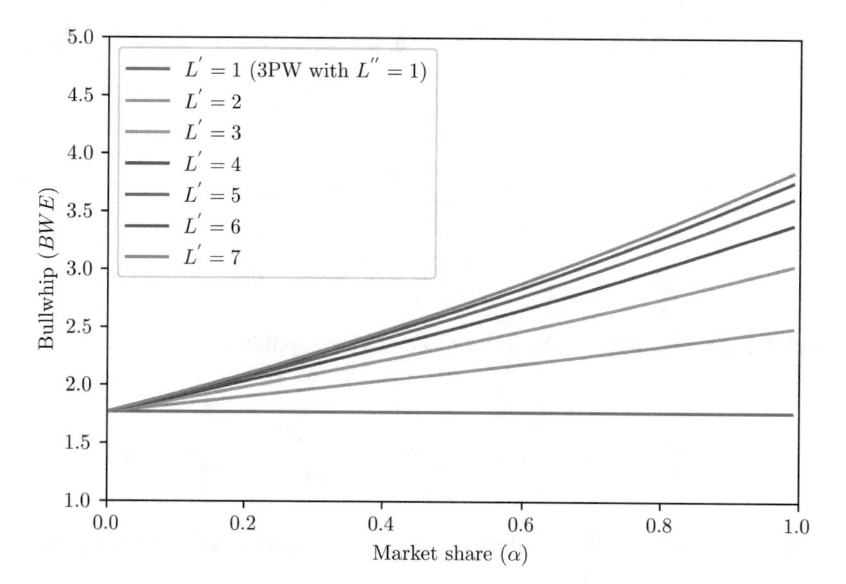

Fig. 8.5 BWE depending on α for $L'' = 1$ and $\rho = 0.6$

8.3.1 Non-MMSE Estimations in a Competing Market

Up to now, we have established that, in the case of iid customer demand, it was irrelevant whether the supplier faces a single retailer facing all customer demand, or two retailers sharing a fixed amount of the customer demand. To reach this result, we assumed that the retailers are able to perform an MMSE estimation of the demand. In this section, we will remove this hypothesis and we will see that this results in a different behaviour of the bullwhip effect.

More specifically, here we consider the case of two retailers and assume an iid customer demand which is forecasted by each retailer using a moving average of parameters $m^{'}$ and $m^{''}$, respectively. Apart from this, we maintain the same hypotheses than in the previous section, i.e. negative returns, OUT replenishment policy, different constant lead times denoted by $L^{'}$ and $L^{''}$, etc. In this case, $\hat{d}_t^{L^{'}}$ the forecast of the demand faced by Retailer #1 is

$$\hat{d}_t^{L^{'}} = \frac{L^{'}}{m^{'}} \sum_{j=0}^{L^{'}} D^{'}_{t-j} \tag{8.44}$$

Therefore, for Retailer #1 the corresponding expression of the order given in Eq. (8.32) is now (take into account that $D^{'}_t = \alpha D_t$)

$$O^{'}_t = \hat{d}_t^{L^{'}} - \hat{d}_{t-1}^{L^{'}} + D^{'}_t = \frac{L^{'}}{m^{'}} \left(D^{'}_t - D_{t-m^{'}+1^{'}} \right) + D^{'}_t = \tag{8.45}$$

$$\alpha \left(1 + \frac{L^{'}}{m^{'}} \right) D_t - \frac{\alpha L^{'}}{m^{'}} D_{t-m^{'}+1} \tag{8.46}$$

Similarly, for Retailer #2 the order placed to the supplier would be

$$O^{''}_t = (1-\alpha) \left(1 + \frac{L^{''}}{m^{''}} \right) D_t - \frac{(1-\alpha) L^{''}}{m^{''}} D_{t-m^{''}+1} \tag{8.47}$$

The sum of the two orders would be O_t the total demand faced by the supplier, i.e.

$$O_t = \alpha \left(1 + \frac{L^{'}}{m^{'}} \right) D_t - \alpha \frac{L^{'}}{m^{'}} D_{t-m^{'}+1} + \tag{8.48}$$

$$(1-\alpha) \left(1 + \frac{L^{''}}{m^{''}} \right) D_t - (1-\alpha) \frac{L^{''}}{m^{''}} D_{t-m^{''}+1} \tag{8.49}$$

Taking variances, and taking into account that the demand is iid

$$V[O_t] = \alpha^2 \left[\left(1 + \frac{L'}{m'}\right)^2 + \left(\frac{L'}{m'}\right)^2 \right] \sigma^2 + \quad (8.50)$$

$$(1-\alpha)^2 \left[\left(1 + \frac{L''}{m''}\right)^2 + \left(\frac{L''}{m''}\right)^2 \right] \sigma^2 = \quad (8.51)$$

$$\alpha^2 \left[1 + 2\frac{L'}{m'} + 2\left(\frac{L'}{m'}\right)^2 \right] \sigma^2 + (1-\alpha)^2 \left[1 + 2\frac{L''}{m''} + 2\left(\frac{L''}{m''}\right)^2 \right] \sigma^2 \quad (8.52)$$

As a consequence, the bullwhip effect can be expressed as

$$BWE = \alpha^2 \left[1 + 2\frac{L'}{m'} + 2\left(\frac{L'}{m'}\right)^2 \right] + (1-\alpha)^2 \left[1 + 2\frac{L''}{m''} + 2\left(\frac{L''}{m''}\right)^2 \right] \quad (8.53)$$

As it can be seen, the bullwhip effect exists even if both lead times and the parameters of the moving average forecasting technique are equal. In this special case, we can see that the bullwhip effect does not depend on α. For the rest of the cases, there is an interaction between the different factors involved (α, the lead times and the number of periods considered in the MA).

8.4 Conclusions

As we have seen, in complex SCs there are different factors usually augmenting the bullwhip effect

- The replenishment cycle of the number of retailers. Even if all retailers are identical, unless their orders are perfectly balanced (so, on average, the supplier can expect the same volume of orders in each period), there is an increase in the bullwhip effect. Such effect is exacerbated by the number of retailers unless they are independent.
- Even if batching ordering does not occur, introducing different lead times, different forecasting methods, or different replenishment policies almost always results in an additional bullwhip effect.

In view of the results obtained, we can see that linear approximation of a non-linear SC structure are heavily influenced by the lack of *balance* among the retailers, as well as their lack of *homogeneity*. As a rule, the farther we move from an echelon-balanced and echelon-homogeneous SC, the more we are underestimating the dynamics effects in the SC.

8.5 Further Readings

Since most of the early models in SC literature usually assume a linear structure and, taking into account that most real-life SCs are non-linear, a recurring research question is whether the bullwhip effect would be higher in the latter settings or not. While, in general, most of the contributions indicate the amplification of the variability of orders and inventory in these more complex settings, some mitigation can be achieved via risk pooling (see e.g. [1]). The case of several retailers with order batching is first described in [2], being pointed out as one of the four causes of the bullwhip effect identified.

Under the hypotheses of linear, time-invariant inventory management policies and stationary customer demand, exact formulae—even if rather complex—for the variance of the orders are given in [3]. The analysis of two retailers with a fixed market shared first appeared in [4]. Other studies of the bullwhip effect in an SC involving several retailers include [5, 6]. The impact of the forecast employed on the value of information sharing in a network with multiple suppliers is studied in [7]. Other work investigating the bullwhip effect in an SC model with one supplier and two retailers using different demand assumptions is [8]. In [9], the bullwhip effect under a dual sourcing environment for an AR(1) demand process is examined analytically

There are many studies addressing the dynamic effects in structurally complex SCs. Given the nature of these settings, most of these references use simulation to study the topic, although some attempts have been made to extract analytical results [10]. Among the simulation studies, we can cite [11, 12] or [13] using agent-based simulation, a modelling methodology that seems quite apt to capture the complexity of real-life SCs.

References

1. Sucky, E.: The bullwhip effect in supply chains–an overestimated problem? Int. J. Prod. Econ. **118**(1), 311–322 (2009). Special Section on Problems and models of inventories selected papers of the fourteenth International symposium on inventories
2. Lee, H., Padmanabhan, V., Whang, S.: Information distortion in a supply chain: the bullwhip effect. Manag. Sci. **43**(4), 546–558 (1997)
3. Ouyang, Y., Li, X.: The bullwhip effect in supply chain networks. Eur. J. Oper. Res. **201**(3), 799–810 (2010)
4. Duc, T.T.H., Luong, H.T., Kim, Y.D.: Effect of the third-party warehouse on bullwhip effect and inventory cost in supply chains. Int. J. Prod. Econ. **124**(2), 395–407 (2010)
5. Khosroshahi, H., Husseini, S., Marjani, M.: The bullwhip effect in a 3-stage supply chain considering multiple retailers using a moving average method for demand forecasting. Appl. Math. Model. **40**(21–22), 8934–8951 (2016)
6. Ma, J., Zhu, L., Yuan, Y., Hou, S.: Study of the bullwhip effect in a multistage supply chain with callback structure considering two retailers. Complexity **2018**,(2018)
7. Zhao, X., Xie, J., Leung, J.: The impact of forecasting model selection on the value of information sharing in a supply chain. Eur. J. Oper. Res. **142**(2), 321–344 (2002)

8. Sirikasemsuk, K., Luong, H.T.: Measure of bullwhip effect in supply chains with first-order bivariate vector autoregression time-series demand model. Comput. Oper. Res. **78**, 59–79 (2017)
9. Sirikasemsuk, K., Luong, H.: Measure of bullwhip effect - a dual sourcing model. Int. J. Oper. Res. **20**(4), 396–426 (2014)
10. Ignaciuk, P., Dziomdziora, A.: Quantifying the bullwhip effect in networked structures with nontrivial topologies, pp. 62–66 (2020)
11. Dominguez, R., Framinan, J., Cannella, S.: Serial vs. divergent supply chain networks: A comparative analysis of the bullwhip effect. Int. J. Prod. Res. **52**(7), 2194–2210 (2014)
12. Dominguez, R., Cannella, S., Framinan, J.: The impact of the supply chain structure on bullwhip effect. Appl. Math. Model. **39**(23–24), 7309–7325 (2014)
13. Dominguez, R., Cannella, S., Framinan, J.: On bullwhip-limiting strategies in divergent supply chain networks. Comput. Ind. Eng. **73**(1), 85–95 (2014)

Chapter 9
Further Issues in Modelling SC Dynamics

9.1 Introduction

Although in the previous chapters, we have discussed many aspects related to the bullwhip effect in SCs, we have also seen that these hardly cover the existing state-of-the-art field and a number of further readings have been suggested at the end of each chapter so the reader can get a more comprehensive view of the last developments. Instead, in this chapter, we focus on some additional areas within the SC dynamics domain that are receiving increasing attention in the last years, and therefore, constitute topics to keep in mind. These areas have been chosen by trying to integrate diverse trends detected in previous reviews on the field, and in view of the number of contributions appearing on these topics.

This chapter is quite different from the previous ones in the book as it does not include models in order to derive insights for certain aspects of SC dynamics. Instead, some of the main trends in the SC dynamics literature are briefly discussed and selected references are presented, even if the literature on these topics is growing very fast and it is not possible to give a full structured review. Instead, this chapter can be read as an extended 'further readings' section on a number of topics, including the following:

- Integration of financial considerations in SC dynamics (Sect. 9.2)
- Enhanced forms of collaboration in the SC (Sect. 9.3)
- Industry/Logistics 4.0: New trends in manufacturing and logistics and how these may impact SC dynamics (Sect. 9.4)
- SC resilience and the ripple effect (Sect. 9.5).

9.2 Financial Considerations in SC Dynamics

Although almost unanimously the literature assumes that the reduction of bullwhip effect results in a better performance of the companies in the SC, there is some

J. M. Framinan, *Modelling Supply Chain Dynamics*,
https://doi.org/10.1007/978-3-030-79189-6_9

evidence showing that this relation is not straightforward; see [1]. In this vein, [2] develop a model that shows that a lower bullwhip effect does not always lead to lower costs in the SC. These results might suggest that there are additional performance aspects that may not be fully captured by using inventory or order variability as surrogate costs and that perhaps require the explicit integration of financial aspects into the existing models to provide more effective insights regarding SC dynamics.

The integration of the financial flow in SC dynamics shows that additional phenomena may occur: pricing decisions taken by independent supply chain nodes may give rise to an amplified or absorbed fluctuation in prices which is referred to as the Bullwhip effect in Pricing (BP) (see e.g. [3]). If the price fluctuation is amplified downstream the SC, then it is referred to as *reverse* BP, whereas, if the price fluctuation is absorbed downstream, then it is denoted as *forward* BP. Several papers address BP and its causes, including [4] or [5]. At a macro level, empirical evidences linking liquidity and bullwhip effect are discussed in [6].

The price of the product for the final customer and its potential to change the demand pattern can also provide interesting insights: In [7], the bullwhip effect is studied in a system with multiple price-sensitive demand streams. In this manner, the individual (at demand stream level) and aggregated conditions to reduce the bullwhip effect can be established. The question of whether the bullwhip effect is higher or lower if the price is considered is answered by [8], who show that both scenarios are possible depending on the magnitude of both lead times and the autocorrelation coefficient of the demand. When pricing considerations are included in the operational parameters of the node/SC (in the same manner as, e.g. replenishment policies or forecasting techniques), the impact of different pricing strategies in the SC dynamics can be studied. Several contributions address these aspects: The joint effect of pricing decisions together with the already-known bullwhip-increasing factors (such as lead times) is studied by [9], while the variability of final customer orders when a supply disruption occur under different pricing strategies is investigated by [10].

Competition among SCs for a price-sensitive final customer also influences the demand pattern in the SCs, and therefore, may affect its dynamic behaviour. In [11], two parallel, two-echelon SCs compete for the final customer demand for a single product, which depends on the prices imposed by each one of the retailers. In this topic; see also [12]. The bullwhip effect in this scenario is measured and the conditions for its reduction or amplification are given. Similar research is carried out by [13] in the context of a CLSC and a non-AR(1) demand pattern. Other papers investigating the bullwhip in the context of price-sensitive demand include [14–16].

Another interesting topic that can be integrated is the consideration of coordination contracts. Although it is usually not considered in most literature addressing SC dynamics, if the financial flow among partners does not correspond to the wholesale contract (i.e. there is a unit price/cost for the product), then the type of contract employed can be used to improve SC dynamics in addition to achieve better coordination. In this regard, the bullwhip effect under revenue-sharing contracts is addressed in [17].

Furthermore, it is to note that the concept of bullwhip effect as the amplification of the variability in the material flow can be extended to other types of flow, including the flow of money in the SC. Particularly, in terms of cash flow, several measures of a *financial* bullwhip effect have been proposed by [18] or [19]. Particularly, the term *bullwhip cash flow* has been introduced by the first authors to name the variability of the cash conversion cycle caused by the bullwhip effect (or more precisely, by the inventory variability), which restricts the liquidity of SC members. These authors also found that the lead times were the main factor influencing this type of bullwhip. In [20], a simulation-optimisation model is developed to reduce the cash flow bullwhip, whereas in [21], it is reviewed how the digitalisation of SCs (addressed in Sect. 9.4 can be used to reduce the bullwhip cash flow). A further reference dealing with the cash flow bullwhip is [22].

9.3 Enhanced Forms of SC Collaboration

SCs are continuously evolving towards new forms of collaboration among the nodes. Many of these forms are associated with the circular economy paradigm, which can be incorporated in the traditional SCM (see in this regard [23] with a review and a discussion of future opportunities). Although, as discussed in Chap. 7, recycling and remanufacturing are being addressed rather intensively in the SC literature, other environmental issues, such as pollution, carbon emission, etc. have received substantially less attention. In particular, the concept of industrial symbiosis [24] (where traditionally separate industries adopt a collective approach to gain competitive advantage by exchanging materials, energy, water and/or by-products), is given rise to new forms of inter-organisational collaboration, most of then following a different, systemic, approach (see e.g. [25]) as opposed to the CLSC paradigm discussed in Chap. 7. Among these, the *symbiotic* SCs can be mentioned as an approach to implement industrial symbiosis (see e.g. [26] or [27]). In these SCs, the flow of materials (and consequently the flow of information and money) does not only occur from one echelon to another, but among nodes belonging to the same echelon, or to echelons in complementary (or symbiotic) SCs. Although the literature in the topic is rapidly growing, most of the papers focus on strategic issues and many tactical or operational aspects (where the dynamics aspects are more stringent) are yet to be understood.

Another trend observed in many sectors is the collaboration of manufacturers and service providers to offer a product product-service (a solution), thus creating the so-called Hybrid SC or HSC. In HSCs, the interaction between inventory/production capabilities and service capacity has a great impact on SC dynamics, as it can be considered that, in addition to the traditional flows (product, information, inventory), a service flow has also to be considered. For this type of SC, [28] develop specific performance metrics for the bullwhip effect and investigate the factors influencing this indicator. This topic is also explored by [29] or [30], among others.

The use of different channels or distributions systems in the SC also may impact the bullwhip effect, as it is shown in [31]. The use of these different transportation modes also opens possibilities for devising suitable replenishment decisions, as in [32].

Furthermore, the new forms of collaboration in an SC also may include the final customers, who can share their product's evaluation online and thus can modify the demand pattern, thus introducing new dynamics in the SC. In this regard, [33] use a simulation model to conclude that an online review system increases both the order and inventory variance amplification and that this impact can be moderated by some attributes such as product quality, unit mismatch cost, lead time and customer volatility.

9.4 Supply Chain 4.0

For many years, the advances and new technologies in manufacturing and logistics had been taken place steadily, but in an incremental fashion. More recently, however, we have witnessed a number of technological developments that are profoundly changing production and logistics. These technologies, sometimes presented under the umbrella of Industry 4.0/Logistics 4.0,[1] include CPS (Cyber-Physical System), IoT (Internet of Things), AI (Artificial Intelligence), blockchain, cloud computing, AM (Additive Manufacturing), among many others. Clearly, these changes impact SC operations as they alter the way in which the material, information and financial flows take place, therefore, substantially altering the dynamics of the SC. The term *Supply Chain 4.0* has been recently proposed by [34] to denote the application of these technologies in the SC context, being [35] a recent review of the contributions in this area. An alternative term employed is that of *Digital Supply Chain* (see [36] or [37]).

Given the extent and depth of these changes, it is not surprising that the number of contributions addressing these issues is steadily growing. Restricting the search to those references addressing specifically system dynamics, the impact of some of these digital technologies (including CPS, Big Data and AI) in the bullwhip effect is analysed in several references, including [38, 39], whereas the use of these *optimised* ordering decisions and/or forecasting decisions (using, e.g. Machine Learning or neural networks) to mitigate the bullwhip effect is addressed in [40–43] or [44]. In [45, 46], the impact of the use of Big Data to reduce the bullwhip effect is studied. The role of IoT in SCM is discussed in [47] or in [48].

The developments in cloud computing can lead to closer collaboration by introducing communication, safety as well as confidentiality standards. Therefore, information inconsistencies leading to the increase of the bullwhip effect such as the IRI discussed in Chap. 6 in supply chains can be potentially reduced and, at the same time, it can increase the trust among the members of the SC. However, the increased communication also poses a number of technical challenges, including security issues [49]. The application of cloud computing in SCs is discussed in [50] and in [38].

AM is another technology that might enormously change many aspects in SCs and in SCM as it may eliminate the need to maintaining inventories for some prod-

[1] In other occasions, the terms *digitalisation* or *digital transformation* are used.

ucts, particularly those with low or intermittent demand, such as spare parts. The theoretical and practical implications of AM in SC are discussed in [51]. Its impact is discussed in [52], whereas in [53], a simulation model is developed for a specific case. The use of AM (Additive Manufacturing) in SCs and its influence in the bullwhip effect is addressed in [54].

Several papers discuss the use of blockchain in the context of SC dynamics. In [55–57], the potential of the blockchain technology to improve information sharing among nodes in a complex SC and is ability to reduce the bullwhip effect is discussed, whereas in [58] and in [59] its specific application to VMI is discussed. The literature on the topic is summarised and discussed in [60] and in [61]. Another paper analysing blockchain in SCs, this time from a transaction cost theory perspective, is [62].

9.5 Supply Chain Resilience and the Ripple Effect

Over the years, SCs have been developed to achieve competitiveness through high customer service, concentrating on their core competencies and improving their operations. In many cases, this fact has increased the interconnection (and therefore, the dependability) among the members of the SC who, in turn, try to operate in a cost-effective manner by removing the excess of production, inventories and idle times. By doing so, their capacity, time and inventory buffers have been substantially reduced, and therefore, the SC is more prone to suffer disruptions due to unexpected events (including natural disasters such as earthquakes, flooding, etc., epidemiological outbreaks such as COVID-19, or financial or economic crises, among others), which can be then enormously damaging (see e.g. [63] for a quantification) and that can even threaten the very existence of the SC [64]. Not surprisingly, many companies are more and more concerned about the capability of their SC to recover from these events [65] or SC *resilience*. Clearly, an important issue is to choose an appropriate level of resilience in the SC, as clearly a high resilience would translate in extra costs due to over-capacity or redundancy. Although perhaps it is an oversimplification, there is a trade-off between the efficiency of the operation of an SC in *normal* conditions, and its resilience (see e.g. [66]). Suitable models would be necessary to help choosing this level of resilience, also taking into account that, although some disruptions can be mistaken by unique events, they are in fact *supply chain tsunamis* in the sense that, as their ecological counterpart, they constitute a recurring phenomenon occurring at very long time intervals [67]. From the SC dynamics perspective, the challenge is to better understand how these unexpected events affect the dynamics behaviour of the nodes and, at the same time, how this behaviour is affected by the structural and operational changes taking place in order to build SC resilience. In this regard, in [68], it is shown that the use of adaptive order policies to mitigate disruptions in the SC can exacerbate the bullwhip effect, hurting the overall performance of the SC. Evidence of a similar trade-off between information management strategies aimed to mitigate disruptions and bullwhip is shown in [69]. The effect of sharing information among SC members (a practice that, as we know, is employed to mitigate the bullwhip effect) might also serve to mitigate disruptions according to [70], although the benefits of info sharing are not equally distributed among the SC members.

A paper analysing the impact of COVID-19 in SC dynamics is [71]. A recent study of the effect of the network structure in the SC resilience is [72], whereas the influence of lead time in SC resilience is studied in [73]. Finally note that, although in line with most of the literature, we have defined SC resilience as the ability of the SC to return to its original state after being disturbed, it is to note that there are different views and definitions of resilience (see e.g. [74]), and some authors also include the ability of the SC to adapt and transform [75]. In this view, additional factors influencing resilience, as well as its dynamic behaviour, must be analysed.

A phenomenon related to the aforementioned unexpected events is the so-called *ripple effect*. The ripple effect occurs when a sudden disruption in the flow of materials in one or a few nodes cannot be contained and it cascades impacting the performance of the entire SC. Some examples of cases of ripple effect in SC have been documented in [76]. The ripple effect is a stressor of the SC resilience, and it is quite different from the bullwhip effect, as the latter refers to high-frequency, low-impact operational risks, whereas the former refers to low (or very low) frequency, but high-impact disturbance risks. Furthermore, the ripple effect is inherently a systemic effect, in contrast to the bullwhip effect, that may be studied in extremely simple settings (i.e. a two-node, serial SC as we have done in Chap. 4). In this regard, it would be interesting to develop models that can reconcile both effects. Some contributions, such as [77–80] explore this issue, although there is, undoubtedly, a need for additional research. It is also interesting to study whether other trends—already discussed in this chapter—such as digitalisation or sustainable SCs are more suitable to increase SC resilience. These aspects are addressed, e.g. in [81] or in [82]. Some additional research directions in this topic are pointed out in [76].

9.6 Conclusions

In this chapter, we have presented four areas within the SC field that are receiving increasing attention as their results can have a great impact in shaping the future SCs. The integration of financial considerations could help to better understand the impact in SC dynamics in the bottom line of the companies, making more evident and accountable phenomena such as the bullwhip effect. Besides, emerging forms of collaboration among companies—some of them linked to the concept of circular economy—foresee different exchanges of materials, information and money that have to be studied in terms of their dynamic behaviour. The ubiquitous introduction of digital technologies under the umbrella of Industry 4.0/Logistics 4.0 also poses a number of challenges in order to be able to reap the benefits brought by these otherwise expensive infrastructures. Finally, at the time of writing these lines, it has been about a year after the world started suffering a pandemic disease that, aside from taking an irreparable toll in lives and changing (hopefully only in a temporary manner) many aspects of our existence, has crippled or damaged many SCs and has put forward the need of increasing their resilience. Of course, in a field so lively and exuberant, it is not possible to know if the interest in these areas will continue in the future, or if other areas will emerge instead. In any case, it does not seem unrealistic to think that the dynamics of the supply chain will continue to be a fruitful research field whose results can be translated to make production and distribution more reliable, efficient and sustainable.

References

1. Mackelprang, A., Malhotra, M.: The impact of bullwhip on supply chains: performance pathways, control mechanisms, and managerial levers. J. Oper. Manag. **36**, 15–32 (2015)
2. Torres, O., Maltz, A.: Understanding the financial consequences of the bullwhip effect in a multi-echelon supply chain. J. Bus. Logist. **31**(1), 23–41 (2010)
3. Ozelkan, E., Çakanyildirim, M.: Reverse bullwhip effect in pricing. Eur. J. Oper. Res. **192**(1), 302–312 (2009)
4. Ozelkan, E., Lim, C.: Conditions of reverse bullwhip effect in pricing for price-sensitive demand functions. Ann. Oper. Res. **164**(1), 211–227 (2008)
5. Adnan, Z., Özelkan, E.: Bullwhip effect in pricing under different supply chain game structures. J. Revenue Pricing Manag. **18**(5), 393–404 (2019)
6. Udenio, M., Fransoo, J., Peels, R.: Destocking, the bullwhip effect, and the credit crisis: empirical modeling of supply chain dynamics. Int. J. Prod. Econ. **160**, 34–46 (2015)
7. Zhang, X., Burke, G.J.: Analysis of compound bullwhip effect causes. Eur. J. Oper. Res. **210**(3), 514–526 (2011)
8. Tai, P., Duc, T., Buddhakulsomsiri, J.: Measure of bullwhip effect in supply chain with price-sensitive and correlated demand. Comput. Ind. Eng. **127**, 408–419 (2019)
9. Gamasaee, R., Zarandi, M.: Incorporating demand, orders, lead time, and pricing decisions for reducing bullwhip effect in supply chains. Sci. Iran. **25**(3E), 1724–1749 (2018)
10. Feng, X., Rong, Y., Shen, Z.J., Snyder, L.: Pricing during disruptions: order variability versus profit. Decision Sciences (2020)
11. Ma, Y., Wang, N., He, Z., Lu, J., Liang, H.: Analysis of the bullwhip effect in two parallel supply chains with interacting price-sensitive demands. Eur. J. Oper. Res. **243**(3), 815–825 (2015)
12. Ma, J., Ma, X.: Measure of the bullwhip effect considering the market competition between two retailers. Int. J. Prod. Res. **55**(2), 313–326 (2017)
13. Giri, B.: Measure of bullwhip effect in a closed-loop supply chain with two retailers under price-sensitive non-arma demand process, pp. 2438–2447 (2020)
14. Wang, N., Lu, J., Feng, G., Ma, Y., Liang, H.: The bullwhip effect on inventory under different information sharing settings based on price-sensitive demand. Int. J. Prod. Res. **54**(13), 4043–4064 (2016)
15. Gao, D., Wang, N., He, Z., Jia, T.: The bullwhip effect in an online retail supply chain: a perspective of price-sensitive demand based on the price discount in e-commerce. IEEE Trans. Eng. Manag. **64**(2), 134–148 (2017)
16. Lu, J., Feng, G., Lai, K., Wang, N.: The bullwhip effect on inventory: a perspective on information quality. Appl. Econ. **49**(24), 2322–2338 (2017)
17. Adnan, Z., Özelkan, E.: Bullwhip effect in pricing under the revenue-sharing contract. Comput. Ind. Eng. **145** (2020)
18. Tangsucheeva, R., Prabhu, V.: Modeling and analysis of cash-flow bullwhip in supply chain. Int. J. Prod. Econ. **145**(1), 431–447 (2013)

19. Chen, T.K., Liao, H.H., Kuo, H.J.: Internal liquidity risk, financial bullwhip effects, and corporate bond yield spreads: supply chain perspectives. J. Bank. Financ. **37**(7), 2434–2456 (2013)
20. Badakhshan, E., Humphreys, P., Maguire, L., McIvor, R.: Using simulation-based system dynamics and genetic algorithms to reduce the cash flow bullwhip in the supply chain. Int. J. Prod. Res. **58**(17), 5253–5279 (2020)
21. Lamzaouek, H., Drissi, H., El Haoud, N.: Digitization of supply chains as a lever for controlling cash flow bullwhip: a systematic literature review. Int. J. Adv. Comput. Sci. Appl. **12**(2), 168–173 (2021)
22. Goodarzi, M., Makvandi, P., Saen, R., Sagheb, M.: What are causes of cash flow bullwhip effect in centralized and decentralized supply chains? Appl. Math. Model. **44**, 640–654 (2017)
23. Lahane, S., Kant, R., Shankar, R.: Circular supply chain management: a state-of-art review and future opportunities. J. Clean. Prod. **258** (2020)
24. Chertow, M.: Industrial symbiosis: literature and taxonomy. Annu. Rev. Energy Environ. **25**, 313–337 (2000)
25. Turken, N., Cannataro, V., Geda, A., Dixit, A.: Nature inspired supply chain solutions: definitions, analogies, and future research directions. Int. J. Prod. Res. **58**(15), 4689–4715 (2020)
26. Herczeg, G., Akkerman, R., Hauschild, M.: Supply chain collaboration in industrial symbiosis networks. J. Clean. Prod. **171**, 1058–1067 (2018)
27. Turken, N., Geda, A.: Supply chain implications of industrial symbiosis: a review and avenues for future research. Resour. Conserv. Recycl. **161** (2020)
28. Lin, W.J., Jiang, Z.B., Liu, R., Wang, L.: The bullwhip effect in hybrid supply chain. Int. J. Prod. Res. **52**(7), 2062–2084 (2014)
29. Wenjin, L., He, R., Qiyun, P., Song, Y., Zhibin, J., Kangzhou, W.: An analysis of the bullwhip effect in multi-echelon hybrid supply chain, pp. 2419–2424 (2019)
30. Dizon, L., Mutuc, J.: Modelling the dynamics of inventory and backlog management in servitized supply chains: the case of custom industrial manufacturing girms, pp. 256–261 (2019)
31. Kadivar, M., Akbarpour Shirazi, M.: Analyzing the behavior of the bullwhip effect considering different distribution systems. Appl. Math. Model. **59**, 319–340 (2018)
32. Keshari, A., Mishra, N., Shukla, N., McGuire, S., Khorana, S.: Multiple order-up-to policy for mitigating bullwhip effect in supply chain network. Ann. Oper. Res. **269**(1–2), 361–386 (2018)
33. Huang, S., Potter, A., Eyers, D.: Using simulation to explore the influence of online reviews on supply chain dynamics. Comput. Ind. Eng. **151**, 106,925 (2021)
34. Frederico, G., Garza-Reyes, J., Anosike, A., Kumar, V.: Supply chain 4.0: concepts, maturity and research agenda. Supply Chain Manag. **25**(2), 262–282 (2019)
35. Barata, J.: The fourth industrial revolution of supply chains: a tertiary study. J. Eng. Technol. Manag. - JET-M **60** (2021)
36. Buyukozkan, G., Gocer, F.: Digital supply chain: literature review and a proposed framework for future research. Comput. Ind. **97**, 157–177 (2018)
37. Ageron, B., Bentahar, O., Gunasekaran, A.: Digital supply chain: challenges and future directions. Supply Chain Forum **21**(3), 133–138 (2020)
38. Wiedenmann, M., Größler, A.: The impact of digital technologies on operational causes of the bullwhip effect – a literature review. Procedia CIRP **81**, 552–557 (2019). 52nd CIRP Conference on Manufacturing Systems (CMS), Ljubljana, Slovenia, June 12–14 (2019)
39. Ran, W., Wang, Y., Yang, L., Liu, S.: Coordination mechanism of supply chain considering the bullwhip effect under digital technologies. Math. Problems Eng. **2020** (2020)
40. Zarandi, M., Pourakbar, M., Turksen, I.: A fuzzy agent-based model for reduction of bullwhip effect in supply chain systems. Expert Syst. Appl. **34**(3), 1680–1691 (2008)
41. Jaipuria, S., Mahapatra, S.: An improved demand forecasting method to reduce bullwhip effect in supply chains. Expert Syst. Appl. **41**(5), 2395–2408 (2014)
42. O'Donnell, T., Humphreys, P., McIvor, R., Maguire, L.: Reducing the negative effects of sales promotions in supply chains using genetic algorithms. Expert Syst. Appl. **36**(4), 7827–7837 (2009)
43. Yousefi, M., Yousefi, M., Ferreira, R.: A review on the application of neural networks for decreasing bullwhip effect in supply chain. Int. Rev. Mech. Eng. **9**(5), 438–442 (2015)

44. Priore, P., Ponte, B., Rosillo, R., de la Fuente, D.: Applying machine learning to the dynamic selection of replenishment policies in fast-changing supply chain environments. Int. J. Prod. Res. **57**(11), 3663–3677 (2019)
45. Hofmann, E.: Big data and supply chain decisions: the impact of volume, variety and velocity properties on the bullwhip effect. Int. J. Prod. Res. **55**(17), 5108–5126 (2017)
46. Hofmann, E., Rüsch, M.: Industry 4.0 and the current status as well as future prospects on logistics. Comput. Ind. **89**, 23–34 (2017)
47. Ben-Daya, M., Hassini, E., Bahroun, Z.: Internet of things and supply chain management: a literature review. Int. J. Prod. Res. **57**(15–16), 4719–4742 (2019)
48. Golpira, H., Khan, S., Safaeipour, S.: A review of logistics internet-of-things: current trends and scope for future research. J. Ind. Inf. Integr. **22** (2021)
49. Sobb, T., Turnbull, B., Moustafa, N.: Supply chain 4.0: a survey of cyber security challenges, solutions and future directions. Electronics (Switzerland) **9**(11), 1–31 (2020)
50. Yu, Y., Cao, R., Schniederjans, D.: Cloud computing and its impact on service level: a multi-agent simulation model. Int. J. Prod. Res. **55**(15), 4341–4353 (2017)
51. Holmstrom, J., Holweg, M., Lawson, B., Pil, F., Wagner, S.: The digitalization of operations and supply chain management: theoretical and methodological implications. J. Oper. Manag. **65**(8), 728–734 (2019)
52. Arbabian, M., Wagner, M.: The impact of 3d printing on manufacturer-retailer supply chains. European Journal of Operational Research **285**(2), 538–552 (2020)
53. Xu, X., Rodgers, M., Guo, W.: Hybrid simulation models for spare parts supply chain considering 3d printing capabilities. J. Manuf. Syst. **59**, 272–282 (2021)
54. Eggenberger, T., Oettmeier, K., Hofmann, E.: Additive manufacturing in automotive spare parts supply chains – a conceptual scenario analysis of possible effects. In: M. Meboldt, C. Klahn (eds.) Industrializing Additive Manufacturing - Proceedings of Additive Manufacturing in Products and Applications - AMPA2017, pp. 223–237. Springer International Publishing, Cham (2018)
55. van Engelenburg, S., Janssen, M., Klievink, B.: A blockchain architecture for reducing the bullwhip effect. Lect. Notes Bus. Inf. Process. **319**, 69–82 (2018)
56. Xue, X., Dou, J., Shang, Y.: Blockchain-driven supply chain decentralized operations - information sharing perspective. Bus. Process Manag. J. **27**(1), 184–203 (2020)
57. Rejeb, A., Keogh, J., Simske, S., Stafford, T., Treiblmaier, H.: Potentials of blockchain technologies for supply chain collaboration: a conceptual framework. Int. J. Logist. Manag. (2021)
58. Guggenberger, T., Schweizer, A., Urbach, N.: Improving interorganizational information sharing for vendor managed inventory: toward a decentralized information hub using blockchain technology. IEEE Trans. Eng. Manag. **67**(4), 1074–1085 (2020)
59. Omar, I., Jayaraman, R., Salah, K., Debe, M., Omar, M.: Enhancing vendor managed inventory supply chain operations using blockchain smart contracts. IEEE Access **8**, 182,704–182,719 (2020)
60. Wan, P., Huang, L., Holtskog, H.: Blockchain-enabled information sharing within a supply chain: a systematic literature review. IEEE Access **8**, 49645–49656 (2020)
61. Hogberg, F., Rashid Othman, M., Grose, C.: Blockchain in supply chains and logistics: trends in development, pp. 852–856 (2020)
62. Schmidt, C., Wagner, S.: Blockchain and supply chain relations: a transaction cost theory perspective. J. Purch. Supply Manag. **25**(4) (2019)
63. Hendricks, K., Singhal, V.: Association between supply chain glitches and operating performance. Manag. Sci. **51**(5), 695–711 (2005)
64. Xu, M., Wang, X., Zhao, L.: Predicted supply chain resilience based on structural evolution against random supply disruptions. Int. J. Syst. Sci. Oper. Logist. **1**(2), 105–117 (2014)
65. World Economic Forum: Building resilence in supply chains (2013)
66. Ivanov, D., Sokolov, B., Dolgui, A.: The ripple effect in supply chains: Trade-off "efficiency-flexibility- resilience" in disruption management. Int. J. Prod. Res. **52**(7), 2154–2172 (2014)
67. Akkermans, H., Van Wassenhove, L.: Supply chain tsunamis: research on low-probability, high-impact disruptions. J. Supply Chain Manag. **54**(1), 64–76 (2018)

68. Schmitt, T., Kumar, S., Stecke, K., Glover, F., Ehlen, M.: Mitigating disruptions in a multi-echelon supply chain using adaptive ordering. Omega (U. K.) **68**, 185–198 (2017)
69. Yang, T., Fan, W.: Information management strategies and supply chain performance under demand disruptions. Int. J. Prod. Res. **54**(1), 8–27 (2016)
70. Sarkar, S., Kumar, S.: A behavioral experiment on inventory management with supply chain disruption. Int. J. Prod. Econ. **169**, 169–178 (2015)
71. Ivanov, D.: Exiting the covid-19 pandemic: after-shock risks and avoidance of disruption tails in supply chains. Ann. Oper. Res. (2021)
72. Li, Y., Zobel, C., Seref, O., Chatfield, D.: Network characteristics and supply chain resilience under conditions of risk propagation. Int. J. Prod. Econ. **223** (2020)
73. Chang, W.S., Lin, Y.T.: The effect of lead-time on supply chain resilience performance. Asia Pac. Manag. Rev. **24**(4), 298–309 (2019)
74. Tukamuhabwa, B., Stevenson, M., Busby, J., Zorzini, M.: Supply chain resilience: definition, review and theoretical foundations for further study. Int. J. Prod. Res. **53**(18), 5592–5623 (2015)
75. Wieland, A., Durach, C.: Two perspectives on supply chain resilience. J. Bus. Logist. (2021)
76. Dolgui, A., Ivanov, D.: Ripple effect and supply chain disruption management: new trends and research directions. International Journal of Production Research **59**(1), 102–109 (2021)
77. Ivanov, D.: Supply chain risk management: bullwhip effect and ripple effect. Int. Ser. Oper. Res. Manag. Sci. **265**, 19–44 (2018)
78. Thomas, A., Mahanty, B.: Interrelationship among resilience, robustness, and bullwhip effect in an inventory and order based production control system. Kybernetes **49**(3), 732–752 (2019)
79. Ivanov, D.: 'a blessing in disguise' or 'as if it wasn't hard enough already': reciprocal and aggravate vulnerabilities in the supply chain. Int. J. Prod. Res. **58**(11), 3252–3262 (2020)
80. Dolgui, A., Ivanov, D., Rozhkov, M.: Does the ripple effect influence the bullwhip effect? an integrated analysis of structural and operational dynamics in the supply chain. Int. J. Prod. Res. **58**(5), 1285–1301 (2020)
81. Yadav, S., Luthra, S., Garg, D.: Modelling internet of things (iot)-driven global sustainability in multi-tier agri-food supply chain under natural epidemic outbreaks. Environ. Sci. Pollut. Res. **28**(13), 16633–16654 (2021)
82. Yilmaz, O., Ozcelik, G., Yeni, F.: Ensuring sustainability in the reverse supply chain in case of the ripple effect: a two-stage stochastic optimization model. J. Clean. Prod. **282** (2021)

Appendix A
Useful Calculus Formulae

A.1 Geometric Series

A geometric series is the sum of number of terms that have a constant ratio ρ between successive terms, i.e.

$$\sum_{i=0}^{n} \rho^i = 1 + \rho + \rho^2 + \cdots + \rho^n \tag{A.1}$$

It is easy to show that this sum is

$$\sum_{i=0}^{n} \rho^i = \frac{1 - \rho^{n+1}}{1 - \rho} \tag{A.2}$$

If the first term of the series is ρ, the following manipulation can be done

$$\sum_{i=1}^{n} \rho^i = \rho \sum_{i=0}^{n-1} \rho^i \tag{A.3}$$

and therefore

$$\sum_{i=1}^{n} \rho^i = \frac{\rho - \rho^{n+1}}{1 - \rho} \tag{A.4}$$

When the number of terms in the summation is infinite ($n \to \infty$) the sum is finite if and only if $|\rho| < 1$, and in this case we have

$$\sum_{i=0}^{\infty} \rho^i = \frac{1}{1 - \rho} \tag{A.5}$$

J. M. Framinan, *Modelling Supply Chain Dynamics*,
https://doi.org/10.1007/978-3-030-79189-6

A.2 Power Series

A power series takes the following form

$$\sum_{i=0}^{n} i^r \cdot \rho^i \tag{A.6}$$

for $r = 1$ we have

$$\sum_{i=0}^{n} i \cdot \rho^i = \rho + 2\rho^2 + \cdots + n\rho^n \tag{A.7}$$

and it is possible to show that such sum is

$$\sum_{i=0}^{n} i \cdot \rho^i = \rho \frac{1-\rho^n}{(1-\rho)^2} - \frac{n \cdot \rho^{n+1}}{(1-\rho)} \tag{A.8}$$

Again, this sum converges for $n \to \infty$ if $|\rho| < 1$:

$$\sum_{i=0}^{\infty} i \cdot \rho^i = \frac{\rho}{(1-\rho)^2} \tag{A.9}$$

A.3 Leibniz's Rule

The general form of Leibniz's rule is the following:

$$\frac{\partial}{\partial x} \int_{a(x)}^{b(x)} f(x, y) dy = \tag{A.10}$$

$$\int_{a(x)}^{b(x)} \frac{\partial f(x, y)}{\partial x} dy + f(x, b(x)) \frac{\partial b(x))}{\partial x} - f(x, a(x)) \frac{\partial a(x))}{\partial x} \tag{A.11}$$

In the book, this formula has been used to solve a special case, i.e. the following expression (see Sect. 2.4):

$$\frac{\partial}{\partial s} \int_{s}^{\infty} x \cdot f(x) dx \tag{A.12}$$

Therefore, the application of the general formula to the special case results in:

$$\frac{\partial}{\partial s} \int_{s}^{\infty} x \cdot f(x) dx = \underbrace{\int_{s}^{\infty} \frac{\partial}{\partial s} (x \cdot f(x)) \, dx}_{\to 0} + 0 - s \cdot f(s) = -s \cdot f(s) \tag{A.13}$$

Appendix B
Basic Probability Tools

In this appendix, we give an overview of the probability tools required to handle the contents of the book. We try to keep it as short as possible, therefore succinct (and possibly not sufficiently precise) definitions are given. Each topic is developed with different depths depending on its use in the book.

B.1 Random Variables

Informally, a (real-valued) random variable or RV is a function whose values depend on the outcomes of a random phenomenon or random experiment. A random experiment is an experiment that can be repeated over time, and where all possible outcomes of the experiment (sample space) are known, even if it is not possible to know the outcome in advance. The fact that it is not possible to know the precise outcome of the experiment does not mean that it has a chaotic behaviour, as some outcomes are more likely than others. Precisely, the probability is a measure to weight the 'likelihood' of an event (a set of outcomes).

RVs are usually denoted with capital letters. Furthermore, the standard notation for the probability apply, i.e.

$$P[a < X \leq b]$$

measures the probability of the event making the RV to have values in the $(a, b]$ interval.

The cumulative distribution function or cdf of a RV is the following function $F_X : \mathbb{R} \rightarrow \mathbb{R}$:

$$F_X(x) = F(x) = P_X((-\infty, x]) = P[X \leq x], \ \forall x \in \mathbb{R}$$

In this book, we dealt either with *discrete* RV or with *continuous* RV. Informally, discrete RVs may take a countable number of values x_i ($i = 1, 2, \ldots$). This implies

J. M. Framinan, *Modelling Supply Chain Dynamics*,
https://doi.org/10.1007/978-3-030-79189-6

that the cdf is a step function (otherwise the number of values of the RV cannot be counted). Then p_i can be defined as the probability that the RV takes the value x_i, i.e. $p_i = P[X = x_i]$, and the collection of the p_i values is denoted as the *probability mass function* or pmf. Clearly, $\sum_i p_i = 1$.

Given the pmf of a discrete RV, its corresponding cdf is easily found by

$$F(x) = P[X \leq x] = \sum_{i \leq x} p_i \tag{B.1}$$

which serves to compute the probability of any event involving this RV.

A continuous RV does not take a countable set of values, which implies that its cdf is continuous and then the probability corresponding to the RV taking a single point value is zero. It is then more appropriate to talk about the probability that the RV is contained in a given interval. In other words, the corresponding probability must be the result of summing across the interval a continuous equivalent of the pmf function. This function is named *probability density function* or pdf and it is denoted by $f(x)$. Therefore, by definition

$$P[a \leq X \leq b] = \int_a^b f(u)du \tag{B.2}$$

From the equation above, it follows that

$$F(x) = P[X \leq x] = \int_{-\infty}^{x} f(u)du \tag{B.3}$$

Equation (B.3) is precisely used sometimes to define $f(x)$ from the fact that $F(x)$ is continuous.

B.1.0.1 Function of a RV

In the informal definition given in Sect. B.1, we defined an RV as a function whose values depended on the outcome of a random experiment. According to this, a function $f(X)$ of an RV X is also an RV, as it would be a *composed* function depending on the outcome of a random experiment.

B.1.1 Main Characteristics of a RV

As we have seen, the cdf of an RV (which, in turn, can be obtained from the pmf or the pdf depending on whether the RV is discrete or continuous) allows computing the likelihood that any event regarding the RV. However, the cdf does not give an intuitive understanding of the behaviour of the RV.

B.1.1.1 Expected Value

The idea of the expected value of a RV is to describe its arithmetic average value after repeating the corresponding random experiment an infinite number of times. For a discrete RV X, given p_i the probability of each outcome x_i, its expected value is denoted as $E[X]$ and is defined as

$$E[X] = \sum_{i=1,2,\ldots} p_i \cdot x_i \tag{B.4}$$

Note that the sum in Eq. (B.4) may not converge. Indeed, it is required that $\sum_{i=1,2,\ldots} p_i \cdot |x_i| < \infty$ to state that $E[X]$ exists, but in the cases within the scope of the book, we will assume that this is always the case.

Also, note that the expected value should not be confused with a *typical* or *most frequent* outcome of the experiment, as it may be that the expected value does not coincide with any possible outcome of the random experiment. Take, for instance, the expected value of an RV representing the outcome of tossing a (regular) dice: Clearly, its expected value (3.5) is not a possible outcome. Instead, the expected value can be seen as a sort of *centre of gravity* of the RV.

The definition of the expected value in the case of a continuous RV can be seen as a translation of Eq. (B.4) to the continuous domain:

$$E[X] = \sum_{-\infty}^{\infty} x \cdot f(x)dx \tag{B.5}$$

The expected value is also denoted as (population) mean, although we will try to avoid this term whenever it leads to confusion with a related, but different, term (sample mean). Commonly, $E[X]$ is also denoted by the Greek letter μ.

Since we know from Sect. B.1 that a function of a RV is another RV, we can write the general expression for the expected value of $g(X)$ a function of the RV X: in the case of a discrete RV, we have:

$$E[g(X)] = \sum_{i=1,2,\ldots} p_i \cdot g(x_i) \tag{B.6}$$

whereas in the case of a continuous RV is:

$$E[g(X)] = \sum_{-\infty}^{\infty} g(x) \cdot f(x)dx \tag{B.7}$$

From Eqs. (B.6) and (B.7), the following properties of the expected value are straightforward:

- $E[c] = c$
- $E[c \cdot X] = c \cdot E[X]$

- $E[g(X) + h(X)] = E[g(X)] + E[h(X)]$

where c is a constant and $g(x)$ and $h(x)$ real-valued functions.

B.1.1.2 Variance of an RV

Clearly, the expected value gives little information of the *randomness* of the RV. We may have two RVs with identical expected value, but in one case, the outcomes are *almost* always close to this expected value, whereas in the other case they are not. We can measure the (quadratic) distance between a value of X and its expected value by constructing the function $g(X) = (X - E[X])^2$. We use the quadratic form to be consistent with the notion of distance from $E[X]$, which is always positive.[1] Since we know that a function of an RV is another RV, we can compute $E[g(X)]$ to measure its expected quadratic distance from the mean. We call this measure *variance* of X and it is denoted as $V[X] = E[(X - E[X])^2]$.

The fact that the variance is the expected value of the quadratic distance implies that it has a different dimension than the RV: if a RV describes the customer demand in terms of product units ordered every day, then the corresponding variance is measured in (product units)^2. Therefore, the *standard deviation* σ can be defined as the positive square root of the variance, i.e.

$$\sigma = \sqrt{V[X]} \tag{B.8}$$

σ is another indicator widely used as it has the same dimension as the RV. Consequently, one can write $V[X] = \sigma^2$.

There are several properties of the variance that can be derived in a straightforward manner taking into account the properties of the expected value in the previous section:

- $V[c] = 0$.
- $V[c \cdot X] = c^2 \cdot V[X]$
- $V[X] = E[(X - E[X])^2] = E[X^2 + E[X]^2 - 2XE[X]] = E[X^2] - E[X]^2$

where c is a constant.

The first property serves to give a practical intuition of the randomness of a real-life phenomenon: if its variance is zero or close to zero, then there is little randomness. Naturally, in many situations, the notion of *close to zero* should be put into context, and one way of doing so is to scale the standard deviation of the RV with respect to its expected value. This indicator is usually denoted as cv, which stands for *coefficient of variation*, and it is defined as

[1]Note that perhaps $g(X) = |X - E[X]|$ could have been used instead, but this less amenable as it is not differentiable for $x = 0$.

$$cv = \frac{\sigma}{E[X]} \tag{B.9}$$

(Note that by using σ we have that cv is dimensionless).

B.2 Random Vectors

Although RVs seem to be sufficiently complicated to handle (at least as compared to its deterministic counterpart), the truth is that, in many situations, we are interested in several characteristics of a given random experiment. Therefore, in general, we can have a collection (vector) of RVs representing different aspects of the random experiment. For instance, we can define the experiment of tossing two dices at the same time and be interested in the values obtained by each dice (each one being represented by an RV), or be interested in the sum of the values obtained (this would be represented by an RV) and, say, the maximum value obtained.[2]

While a random vector is not a random variable, it is clear that a **function** of a random vector is, indeed a random variable: if we have $g : \mathbb{R}^2 \rightarrow \mathbb{R}$, then it is clear that $g(X, Y)$ is an RV. Although perhaps not immediately clear, an example can be seen can be given by the random experiment of the two dices: it is obvious that the sum of the values obtained by the two dices (each one a RV, and therefore, a function of the two RVs) is an RV.

In general, we can define a collection of n RVs, but since this further complicates the notation of the subsequent definitions and proofs, in most of the discussion we will consider the case of two random variables X and Y (bidimensional random vector) and leave to the reader the generalisation for the n-dimensional case.

As an extension to the cdf in RVs, we can define a joint cumulative distribution function, which is the following function $F : \mathbb{R}^2 \rightarrow \mathbb{R}$:

$$F(x, y) = P[X \leq x; Y \leq y] \tag{B.10}$$

The joint cumulative distribution function expresses the probability that RV X takes a value less than or equal to x and, simultaneously, RV Y takes a value less than or equal to y.

If both X and Y are discrete RVs, then we know that X takes values in a countable set of values x_i whereas Y takes values in a countable set of values y_l. Therefore, a joint pmf can be defined as

$$p_{kl} = P[X = x_k; Y = y_l] \tag{B.11}$$

[2]Although we will formalise the concept of independent RVs later on, intuitively we see that in the first case the chosen variables are independent—informally, their values do not depend one on another—, while in the second case it is clear that the two RV are "related". Indeed, in many practical applications, the interest is in describing such a relationship.

p_{kl} expresses the probability that X takes the value x_k and Y the value y_l.

Similarly, for the case where both X and Y are continuous RVs, the joint pdf can be defined as the function $f(x, y)$ that verifies

$$F(x, y) = P[X \leq x; Y \leq y] = \int_{-\infty}^{x} \int_{-\infty}^{y} f(u, v)dudv \tag{B.12}$$

Given a discrete random vector (X, Y) and the corresponding joint pmf p_{kl}, the *marginal* pmf's (i.e. the pmf's of each one of the RV composing the random vector) are

$$p_k = P[X = x_k] = \sum_{\forall y_l} p_{kl} = \sum_{\forall y_l} P[X = x_k; Y = y_l] \tag{B.13}$$

for RV X, and

$$p_l = P[Y = y_l] = \sum_{\forall x_k} p_{kl} = \sum_{\forall x_k} P[X = x_k; Y = y_l] \tag{B.14}$$

for RV Y.

Similarly, for a continuous random vector (X, Y), given the corresponding joint pdf $f(x, y)$, the *marginal* pdf's are

$$f_X(x) = \int_{-\infty}^{\infty} f(x, y)dy \tag{B.15}$$

and

$$f_Y(y) = \int_{-\infty}^{\infty} f(x, y)dx \tag{B.16}$$

B.2.1 Properties of the Random Vectors

It is not easy to summarise the behaviour of a random vector by defining some characteristics such as those in Sect. B.1.1 for the RVs, as it makes not so much sense to define the *centre of gravity* of a random vector or its variability around this centre of gravity. However, it is possible to define the expected value of a function of a random vector: if we have $g : \mathbb{R}^2 \rightarrow \mathbb{R}$, we have already discussed that $g(X, Y)$ is an RV, and therefore, $E[g(X, Y)]$ can be defined. In the case of a discrete random vector, we have

$$E[g(X, Y)] = \sum_{\forall x_k, y_l} g(x_k, y_k) \cdot p_{kl} \tag{B.17}$$

In the case of a continuous random vector

$$E[g(X, Y)] = \int_{-\infty}^{+\infty} \int_{-\infty}^{+\infty} g(x, y) f(x, y)dxdy \tag{B.18}$$

As with RVs, we assume that the sum and the integral will exist for the cases described in the book. However, this is not the case in general functions.

From the above definitions, it can be seen in a straightforward manner that the following property holds:

$$E[ag(X, Y) + bh(X, Y)] = aE[g(X, Y)] + bE[h(X, Y)] \tag{B.19}$$

with a, b constants and $g, h : \mathbb{R}^2 \to \mathbb{R}$. A particularly interesting case is where $a = b = 1$ and $g(X, Y) = X$, $h(X, Y) = Y$. In this case, we have

$$E[X + Y] = E[X] + E[Y] \tag{B.20}$$

which is usually expressed stating that the expected value of the sum of two RV is the sum of their expected values. The generalisation of this expression for n RVs X_i $(i = 1, \ldots, n)$ is

$$E[\sum_{i=1}^{n} X_i] = \sum_{i=1}^{n} E[X_i] \tag{B.21}$$

It does not make sense to define the variance of a RV, but an interesting indicator can be obtained by defining $Cov[X, Y]$ the **covariance** between two RVs X and Y in the following manner:

$$Cov[X, Y] = E[(X - E[X])(Y - E[Y])] \tag{B.22}$$

As the covariance is the expected value of a RV (a function of a random vector), it is computed in a different way depending on whether the random vector is discrete or continuous. For the discrete case is

$$Cov[X, Y] = \sum_{\forall x_k, y_l} (x_k - E[X])(y_l - E[Y])p_{kl} \tag{B.23}$$

which provides perhaps the easiest manner to interpret the covariance. Let us assume for the moment that p_{kl} are the same across the summation. From the equation, we see that the contribution of one term of the summation would be (positively) high if X yields a high value (with respect to its expected value) and Y yields also a high value (with respect to its expected value), or if both RVs yield low values. Similarly, the contribution would be highly negative if, when one RV is high (low) with respect to its expected value, the other is low (high). Since all these contributions are weighted by the probability that the two values happen simultaneously and then summed, we see that the covariance is measuring the *causal relationship* between the two RVs. For two RVs which are not related at all, we would expect that the positive contributions and the negative contributions would cancel each other, and thus the covariance is zero. A covariance is higher than zero means that the two RVs are directly related, whereas if the covariance is negative, then the two RVs are inversely related.

In the continuous case, the definition of covariance is

$$Cov[X, Y] = \int_{-\infty}^{\infty} \int_{-\infty}^{\infty} (x - E[X])(y - E[Y]) f(x, y) dx dy \tag{B.24}$$

The following properties (which can be proved rather easily) hold for the covariance:

- $Cov[X, X] = V[X]$
- $Cov[X, Y] = E[XY] - E[X]E[Y]$
- $Cov[X, Y] = Cov[Y, X]$
- $V[X + Y] = V[X] + V[Y] + 2Cov[X, Y]$

The last property can be generalised for n RVs $X_1, \ldots, X_n$

$$V\left[\sum_{i=1}^{n} X_i\right] = \sum_{i=1}^{n} V[X_i] + 2\sum_{i<j} Cov[X_i, X_j] \tag{B.25}$$

As it can be seen, the sum of the variance is not, in general, the variance of the sum.

B.2.2 Conditional Probability and Independence

We introduce next the concept of conditioned probability. This concept arises when we have some knowledge regarding one of the RV and we would like to use that knowledge to guess the values of the other RV. In the case of a discrete random vector (X, Y), we can define the conditional pmf for X given that $Y = y_l$ as follows:

$$p_{k|l} = P[X = x_k | Y = y_l] = \frac{P[X = x_k; Y = y_l]}{P[Y = y_l]} = \frac{p_{kl}}{p_l} \tag{B.26}$$

and

$$p_{l|k} = P[Y = y_l | X = x_k] = \frac{P[X = x_k; Y = y_l]}{P[X = x_k]} = \frac{p_{kl}}{p_k} \tag{B.27}$$

As we can see, in this manner, $X|Y$ $(Y|X)$ can be loosely interpreted as an RV with $p_{k|l}$ $(p_{l|k})$ as pmf. As such, we can obtain some characteristics of this RV, such as its expected value or its variance. This aspect will be discussed in the next subsection.

In an analogous manner, in the case of a continuous random vector (X, Y), we can define the conditional pdf for X given that $Y = y$ as follows:

$$f_{X|Y=y}(x) = \frac{f(x, y)}{f_Y(y)} \tag{B.28}$$

and

$$f_{Y|X=x}(y) = \frac{f(x, y)}{f_X(x)} \tag{B.29}$$

As in the discrete case, $X|Y$ $(Y|X)$ can be interpreted as a RV with $f_{X|Y=y}(x)$ $(f_{Y|X=x}(y))$ as pdf.

B.2.3 Law of Total Expectation

Let (X, Y) be a random vector. Then:

$$E[X] = E[E[X|Y]] \tag{B.30}$$

where $E[X|Y]$ is the expected value of X conditioned to Y. If (X, Y) is a discrete random vector, then

$$E[X|Y] = \sum_{\forall x_k} p_{k|l} \cdot x_k = \sum_{\forall x_k} \frac{p_{kl}}{p_l} \cdot x_k \tag{B.31}$$

Note that $E[X|Y]$ can be interpreted as an RV, as it is a function of the random variable Y, so it is possible to obtain its expected value:

$$E[E[X|Y]] = \sum_{\forall y_l} \left(\sum_{\forall x_k} \frac{p_{kl}}{p_l} \cdot x_k \right) \cdot p_l = \sum_{\forall x_k} \left(\sum_{\forall y_l} p_{kl} \right) \cdot x_k = \tag{B.32}$$

$$\sum_{\forall x_k} p_k \cdot x_k = E[X] \tag{B.33}$$

A similar proof can be done if the random vector is formed by continuous variables.

B.2.4 Law of Total Variance

Let (X, Y) be a random vector. Then

$$V[X] = E[V[X|Y]] + V[E[X|Y]] \tag{B.34}$$

The proof is relatively straightforward using the result from Eq. (B.30):

$$V[X] = E[X^2] - E[X]^2 = E[E[X^2|Y]] - E[E[X|Y]]^2 = \tag{B.35}$$

$$E[V[X|Y] + E[X|Y]^2] - E[E[X|Y]]]^2 = \tag{B.36}$$

$$E[V[X|Y]] + E[E[X|Y]^2] - E[E[X|Y]]^2 = \tag{B.37}$$

$$E[V[X|Y]] + V[E[X|Y]] \tag{B.38}$$

B.2.5 Random Sum

Let us assume N a RV, and X_i a random sample following the distribution of X. The RV $Z = \sum_{i=1}^{N} X_i$ is said to be a *random sum*, as the number of elements in the summation is itself a RV. Using the laws of total expectation and total variance in the previous Sections it is possible to derive expressions for $E[Z]$ and $V[Z]$:

$$E[Z] = E[\sum_{i=1}^{N} X_i] = E[E[\sum_{i=1}^{N} X_i|N]] = E[\sum_{i=1}^{N} E[X_i]] = E[NE[X]] = E[N] \cdot E[X] \quad \text{(B.39)}$$

and

$$V[Z] = V[E[\sum_{i=1}^{N} X_i|N]] + E[V[\sum_{i=1}^{N} X_i|N]] = \quad \text{(B.40)}$$

$$V[N \cdot E[X]] + E[N \cdot V[X]] = E[X]^2 \cdot V[N] + V[X] \cdot E[N] \quad \text{(B.41)}$$

B.3 Independence of RVs

Two discrete RVs X and Y with joint pmf p_{kl} and marginal pmfs p_k and p_l respectively, are called independent if

$$p_{kl} = p_k \cdot p_l \quad \forall \; x_k, y_l$$

If X and Y are continuous, the equivalent condition is

$$f(x, y) = f_X(x) \cdot f_Y(y) \quad \forall x, y \in \mathbb{R}$$

being $f_X(x)$ and $f_Y(y)$ the corresponding marginal pdfs.

As a consequence, the following equality holds:

$$E[XY] = E[X] \cdot E[Y]$$

As, since $Cov[X, Y] = E[XY] - E[X] \cdot E[Y]$ it follows that $Cov[X, Y] = 0$, and then the variance of the sum results to be the sum of the variances.

The concept of independence can be generalised to n RVs, and in this case, we say that these RVs are *mutually independent*. Finally, we can say that they are *independent and identically distributed* (iid) if, aside from being mutually independent, they have the same cdf.

B.4 Useful Distributions

In this section, we present some brief descriptions of some distribution functions that are employed in different passages of the book. The list is by no means exhaustive and the reader should not expect a full description, but rather the main concepts that are required to understand the corresponding passages.

B.4.1 Bernoulli

A discrete RV $X \in \{0, 1\}$ is said to have a Bernoulli distribution if its corresponding pmf is

$$p_k = \begin{cases} 1-p & \text{for } k = 0 \\ p & \text{for } k = 1 \end{cases} \tag{B.42}$$

where $0 < p < 1$. As it can be seen, p is the only parameter of the distribution, and it is denoted $X \sim Ber(p)$. This distribution is used to model a random experiment with two possible outcomes (success and failure). The convention is that the success is assigned $k = 1$ (and $k = 0$ to the failure), therefore, p can be interpreted as the success probability of the random experiment.

From the definitions of expected value and variance given in Sect. B.1.1, it follows that $E[X] = p$ and that $V[X] = p \cdot (1 - p)$. Sometimes $q = 1 - p$ is employed to simplify the description.

B.4.2 Binomial

A discrete RV $X \in \{0, n\}$ is said to have a binomial distribution (with parameters n and $p \in (0, 1)$) if its corresponding pmf is

$$p_k = \binom{n}{k} p^k q^{n-k} = \frac{n!}{k!(n-k!)} p^k q^{n-k},\ k = 0, \ldots, n \tag{B.43}$$

The distribution is denoted as $Bi(n, p)$, so $X \sim Bi(n, p)$. This distribution is employed to model a random experiment consisting of performing n independent trials of a Bernoulli-type random experiment (with success probability p), and summing the number of successes (we can check that the pmf evaluated at k gives the probability of k successes out of n trials). In other words, we can write

$$X = \sum_{j=1}^{n} Z_j \tag{B.44}$$

with $Z_j \sim Ber(p)$. Either applying the definition of expected value or variance given in Sect. B.1.1, or simply by applying the properties of the expected value of the sum of RVs, or the variance of the sum of independent RVs, we have $E[X] = np$ and $V[X] = np(1-p)$.

B.4.3 Normal

A continuous RV X has a normal distribution with parameters μ and $\sigma > 0$ if its pdf is

$$f(x) = \frac{1}{\sigma\sqrt{2\pi}} e^{-\frac{(x-\mu)^2}{2\sigma^2}} \quad \text{for } -\infty < x < \infty \tag{B.45}$$

The expected value of X is $E[X] = \mu$ and the variance is $V[X] = \sigma^2$. A special case of the normal distribution where $\mu = 0$ and $\sigma = 1$ is the *standard normal distribution*. An interesting property is that if $X \sim N(\mu, \sigma^2)$, then $\frac{X-\mu}{\sigma} \sim N(0, 1)$, so any normal distribution can be *standardised*.

The normal distribution is the most common distribution appearing in many natural phenomena. The underlying reason is that the sample means of any RV will limit to the normal (central limit theorem).

B.4.4 Gamma

A continuous RV X has a gamma distribution with parameters $\alpha > 0$ and $\beta > 0$ if its pdf is

$$f(x) = \frac{\beta^\alpha}{\Gamma(\alpha)} x^{\alpha-1} e^{-\beta x} \quad \text{for } 0 < x < \infty \tag{B.46}$$

where $\Gamma(\alpha)$ is the gamma function evaluated at point α. The gamma function is defined as

$$\Gamma(s) = \int_0^\infty s^{x-1} e^{-s} dx \tag{B.47}$$

The gamma distribution is typically employed to model times since, in contrast to the normal distribution, a gamma-distributed RV only takes positive values. It can be shown that $E[X] = \frac{\alpha}{\beta}$ and that $V[X] = \frac{\alpha}{\beta^2}$.

Appendix C
Time Series Basics for Demand Modelling

This appendix gives an overview of the minimum time series knowledge required to handle the contents of the book, particularly those related to demand modelling, which is heavily based on the AR(1) model. As with the rest of the appendices, the topic is explained in a rather informal way with the sole objective that the reader can follow the models developed in the book.

C.1 Stationary Time Series

Time series analysis builds upon the theory of stochastic processes, in which we have a sequence of RVs $D_t, D_{t+1}, \ldots$. In general, each RV can follow a different distribution (and consequently different means and variances). However, we will deal here with *stationary* time series. The basic idea of stationarity is that the probability laws that govern the behaviour of the stochastic process do not change over time. In a strict stationary series, the joint distribution of the demand in an consecutive number of time points do not change over time, but we would only require that the demand is weakly stationary, for which it is sufficient to fulfil two conditions:

- The mean of the demand does not change over time, i.e. $E[D_t] = E[D_{t+1}] = E[D] \;\; \forall\, t$
- The covariance between two variables D_t and D_{t+h} $(h = 0, 1, \ldots)$ only depends on h the difference in time between these two demands, and not on t. Note that this last condition implies that the variance of D_t does not depend on t, i.e. $V[D_t] = V[D_{t+1}] = V[D] \;\; \forall\, t$.

Note that the definition of a weakly stationary series does not mean that D_t and D_{t+k} behave the same (are the same random variable).

J. M. Framinan, *Modelling Supply Chain Dynamics*,
https://doi.org/10.1007/978-3-030-79189-6

C.1.1 AR(1) Series

A class of stationary time-series models is the so-called ARMA (AutoRegressive Moving Average) models. These class of models have a great importance in modelling demand. Among them, the most important is the so-called first-order autoregressive model –or AR(1) model for short–, where D_t the demand observed in a period can be expressed as:

$$D_t = d + \rho \cdot D_{t-1} + \epsilon_t \tag{C.1}$$

where $d \geq 0$ is a constant, ρ is the so-called correlation constant which must verify that $|\rho| < 1$, and ϵ_t is assumed to follow a normal distribution with mean zero and variance σ^2, i.e. $\epsilon_t \sim N(0, \sigma^2)$. From this assumption it follows that ϵ_t are iid random variables.

ϵ_t is also known as *white noise*, as it reflects the inherent randomness that cannot be capture by other terms in Eq. (2.1). The correlation constant ρ expresses how the demand in the past periods is related to the demand in the next period. Indeed, Eq. (2.1) can be written recursively as a function of D_{t-2}:

$$D_t = d(1 + \rho) + \rho^2 \cdot D_{t-2} + \epsilon_t + \rho \cdot \epsilon_{t-1}$$

And, applying the recursion and assuming that we have an infinite series in the past, it could be written exclusively in terms of the constants and the white noise

$$D_t = d \sum_{i=0}^{\infty} \rho^i + \sum_{i=0}^{\infty} \rho^i \cdot \epsilon_{t-i} \tag{C.2}$$

Equation (C.2) is sometimes referred to as the *causal form* of the AR(1) model and, in this form, the requirement of $|\rho| < 1$ appears clear, since otherwise the expected value of D_t would not exist as the first term in Eq. (C.2) would not converge (note that the expected value of the second term is zero as we assumed that $E[\epsilon_t] = 0$).

C.1.1.1 Mean, Variance and Covariance

Recall that one condition that we have defined for the stationary series is that its mean and variance are the same over time, i.e. $E[D_t] = E[D_{t+1}] = E[D] \ \forall\, t$, and $V[D_t] = V[D_{t+1}] = V[D] \ \forall\, t$. For the AR(1), we can derive the expression for these values. We will show the following expressions, which will be used intensively in the book:

- $E[D] = \frac{d}{1-\rho}$
- $V[D] = \frac{\sigma^2}{1-\rho^2}$
- $cov(D_t, D_{t-k}) = \rho^k \cdot V[D] = \rho^k \frac{\sigma^2}{1-\rho^2}$

In the remainder of the section, we will obtain these expressions. With respect to the expected value, taking expectations in Eq. (C.1), we have

$$E[D_t] = d + \rho \cdot E[D_{t-1}] + E[\epsilon_t] \tag{C.3}$$

and taking into account the stationary condition of the series and that the expected value of ϵ_t is zero, qe obtain that

$$E[D] = d + \rho \cdot E[D] \tag{C.4}$$

from where we have

$$E[D] = \frac{d}{1-\rho} \tag{C.5}$$

Since usually the mean of the demand is denoted as μ, we have that for an AR(1) demand, we have $\mu = \frac{d}{1-\rho}$.

Regarding the variance, it can be obtained using the same procedure as before, i.e. taking the variance in (C.1)

$$V[D_t] = \rho^2 \cdot V[D_{t-1}] + V[\epsilon_t] \tag{C.6}$$

Note that the right term of the expression can be obtained as D_{t-1} depends on the past demand, and therefore, on $\epsilon_{t-1}, \epsilon_{t-2}, \ldots$ (see Eq. (C.2)), but not on ϵ_t (which aside are iid). Since $V[\epsilon_t] = \sigma^2$, we obtain

$$V[D] = \frac{\sigma^2}{1-\rho^2} \tag{C.7}$$

Regarding the expression of the covariance between two demands separated by k periods, i.e. $cov(D_t, D_{t-k})$, we will first prove that $cov(D_t, D_{t-k}) = \rho \cdot cov(D_{t-1}, D_{t-k})$. To do so, we recall the property of the covariance discussed in Appendix C

$$cov(D_t, D_{t-k}) = E[D_t \cdot D_{t-k}] - E[D_t] \cdot E[D_{t-k}] = E[D_t \cdot D_{t-k}] - E[D]^2 \tag{C.8}$$

The first term in the right can be developed by taking Eq. (C.1) and multiplying both terms of the equation y D_{t-k}, i.e.

$$D_t \cdot D_{t-k} = d \cdot D_{t-k} + \rho \cdot D_{t-1} D_{t-k} + \epsilon_t \cdot D_{t-k} \tag{C.9}$$

Taking expectations in the above equation

$$E[D_t \cdot D_{t-k}] = dE[D] + \rho \cdot E[D_{t-1} D_{t-k}] + E[\epsilon_t \cdot D_{t-k}] \tag{C.10}$$

Note that the last term in the right is zero, since ϵ_t and D_{t-k} are independent (see the reasoning above), therefore, $E[\epsilon_t \cdot D_{t-k}] = E[\epsilon_t] \cdot E[D_{t-k}]$ and $E[\epsilon_t] = 0$. Besides, the term $E[D_{t-1}D_{t-k}]$ can be written as $cov(D_{t-1}, D_{t-k}) + E[D]^2$ (see Eq. (C.8)), therefore, we have

$$E[D_t \cdot D_{t-k}] = d \cdot E[D] + \rho\left(cov(D_{t-1}, D_{t-k}) + E[D]^2\right) \tag{C.11}$$

Plugging this equation in Eq. (C.8), we obtain

$$cov(D_t, D_{t-k}) = d \cdot E[D] + \rho\left(cov(D_{t-1}, D_{t-k}) + E[D]^2\right) - E[D]^2 = \tag{C.12}$$

$$d \cdot E[D] + \rho cov(D_{t-1}, D_{t-k}) - E[D]^2(1-\rho) = \tag{C.13}$$

$$\rho cov(D_{t-1}, D_{t-k}) \tag{C.14}$$

For the case $k = 1$, we would have

$$cov(D_t, D_{t-1}) = \rho cov(D_{t-1}, D_{t-1}) = \rho \cdot V[D] = \rho\frac{\sigma^2}{1-\rho^2} \tag{C.15}$$

Therefore, Eq. (C.12) can be written as

$$cov(D_t, D_{t-k}) = \rho^k \cdot V[D] = \rho^k\frac{\sigma^2}{1-\rho^2} \tag{C.16}$$

C.1.2 ARMA(p, q) Series

As mentioned before, the AR(1) is a special case of the more general stationary ARMA time series. This general model has two parameters: p to indicate the number of autoregression periods, and q to indicate the number of moving average periods considered. More specifically, an ARMA(p,q) model has the following form:

$$D_t = \mu + \underbrace{\phi_1(D_{t-1}-\mu) + \phi_2(D_{t-2}-\mu) + \cdots + \phi_p(D_{t-p}-\mu)}_{\text{Autoregressive part}} + \epsilon_t \tag{C.17}$$

$$\underbrace{-\theta_1\epsilon_{t-1} - \cdots - \theta_q\epsilon_{t-q}}_{\text{Moving average part}} \tag{C.18}$$

where $\mu = E[D]$ and, as usual, $\epsilon_t \sim N(0, \sigma^2)$. As it can be seen, for $p = 1$ and $q = 0$, if $\phi_1 = \rho$, we obtain the AR(1) model already discussed. We will not use the general ARMA model in the book, although it is worth mentioning that it can also be written in a causal form: if we assume that we only have one demand in the series corresponding to $t = 1$, we have

$$D_1 - \mu = \epsilon_1 \tag{C.19}$$

For $t = 2$, we have

$$D_2 - \mu = \phi_1(D_1 - \mu) + \epsilon_2 - \theta_1\epsilon_1 \tag{C.20}$$

and, substituting the expression of D_1

$$D_2 - \mu = \phi_1 \cdot \epsilon_1 + \epsilon_2 - \theta_1\epsilon_1 = \epsilon_2 + (\phi_1 - \theta_1)\epsilon_1 \tag{C.21}$$

For $t = 3$

$$D_3 - \mu = \phi_1(D_2 - \mu) + \phi_2(D_1 - \mu) + \epsilon_3 - \theta_2\epsilon_2 - \theta_1\epsilon_1 \tag{C.22}$$

And, again, substituting the expressions of D_2 and D_1

$$D_3 - \mu = \phi_1(\epsilon_2 + (\phi_1 - \theta_1)\epsilon_1) + \phi_2\epsilon_1 + \epsilon_3 - \theta_2\epsilon_2 - \theta_1\epsilon_1 = \tag{C.23}$$

$$\epsilon_3 + (\phi_1 - \theta_2)\epsilon_2 + [\phi_1(\phi_1 - \theta_1) + \phi_2 - \theta 1]\epsilon_1 \tag{C.24}$$

In this manner, assuming that the series has infinite data, it is easy to check that it is possible to write the series as a linear function of the white noise, i.e.

$$D_t = \mu + \epsilon_t + \psi_1\epsilon_{t-1} + \psi_2\epsilon_{t-2} + \psi_3\epsilon_{t-3} + \dots \tag{C.25}$$

where the coefficients ψ_i $(i = 1, \dots)$ are constant (not RVs). By defining that $\psi_0 = 1$, Eq. (C.25) can be written in a more compact manner

$$D_t = \mu + \sum_{i=0}^{\infty} \psi_i\epsilon_{t-i} \tag{C.26}$$

C.2 Minimum Mean Square Estimation (MMSE)

One interesting question regarding the time series is to estimate what would be the minimum error that one can expect in the long run. For a start, from Sect. 2.3.4, we know different indicators to measure the error, so in this case, we will focus on the MSE indicator. If the set of historical data is sufficiently large, it is clear that MSE would converge to the expected value of the squared difference between the demand—now considered a random variable—and the estimation of the demand, so the following alternative definition of MSE can be used:

$$MSE(\hat{d}_t(h)) = E[(\hat{d}_t(h) - D_{t+h})^2] \tag{C.27}$$

Assuming that the demand can be modelled using an ARMA model, we know that the demand in time period $t + h$ can be written as follows (see Eq. (C.26)):

$$D_{t+h} = \mu + \epsilon_{t+h} + \psi_1 \epsilon_{t+h-1} + \cdots + \psi_{h-1} \epsilon_{t+1} + \tag{C.28}$$
$$+\psi_h \epsilon_t + \psi_{h+1} \epsilon_{t-1} + \ldots \tag{C.29}$$

Note that the interpretation of the two lines is the following: if we are in time period t, the last line in the equation can be interpreted as the 'past' noise (i.e. from time periods before time period t), while the first one can be interpreted as the 'future' noise.

Let us define the following generic estimator applied in time period t to forecast the demand in time period $t + h$

$$\hat{D}_t(h) = \mu + \alpha_h \epsilon_t + \alpha_{h+1} \epsilon_{t-1} + \ldots \tag{C.30}$$

where α_h are parameters to indicate how to weigh the noise from previous periods. Note that the estimator in Eq. (C.30) is generic in the sense that we would try to determine the best values of α_h so the estimator is optimal (according to the MSE indicator). Therefore, we first construct the difference between D_{t+h} and $\hat{D}_t(h)$:

$$D_{t+h} - \hat{D}_t(h) = \epsilon_{t+h} + \psi_1 \epsilon_{t+h-1} + \cdots + \psi_{h-1} \epsilon_{t+1} + \tag{C.31}$$
$$+(\psi_h - \alpha_h)\epsilon_t + (\psi_{h+1} - \alpha_{h+1})\epsilon_{t-1} + \ldots \tag{C.32}$$

The previous equation can be written in a more compact manner, i.e.

$$D_{t+h} - \hat{D}_t(h) = \sum_{i=0}^{h-1} \psi_i \epsilon_{t+h-i} + \sum_{i=h}^{\infty} (\psi_i - \alpha_i)\epsilon_{t-i+h} \tag{C.33}$$

Taking squares on both sides of Eq. (C.33) and then taking the expected value, it can be seen that the term in the right consists on a cross-product of terms $E[\epsilon_i \cdot \epsilon_k]$ for different values of i and k. Since, by definition, ϵ_i are iid, then the expected value of their product is zero unless $i = k$. Otherwise, recall that $E[\epsilon_i^2] = E[\epsilon_i]^2 + V[\epsilon_i] = \sigma^2$. Therefore

$$MSE = E\left[\left(D_{t+h} - \hat{D}_t(h)\right)^2\right] = \sum_{i=0}^{h-1} \psi_i^2 \sigma^2 + \sum_{i=h}^{\infty} (\psi_i - \alpha_i)^2 \sigma^2 \tag{C.34}$$

Clearly, MSE is minimal (Minimal MSE or $MMSE$) if α_i are chosen in such a way so $\phi_i = \alpha_i$ for $i = h, \ldots, \infty$. As a result, the $MMSE$ estimator is simply

$$\hat{D}_t(h) = \mu + \psi_h \epsilon_t + \psi_{h+1} \epsilon_{t-1} + \ldots \tag{C.35}$$

Finally, note that the error—denoted as $e_t(h)$—obtained when using this estimator is

$$e_t(h) = D_{t+h} - \hat{D}_t(h) = \sum_{i=0}^{h-1} \psi_i \epsilon_{t+h-i} \tag{C.36}$$

Index

J. M. Framinan, *Modelling Supply Chain Dynamics*,
https://doi.org/10.1007/978-3-030-79189-6

Printed in the United States
by Baker & Taylor Publisher Services